Vibration Analysis and Structural Dynamics for Civil Engineers

Vibration Analysis and Structural Dynamics for Civil Engineers

Essentials and Group-Theoretic Formulations

Alphose Zingoni

University of Capetown
South Africa

CRC Press

Taylor & Francis Group

Boca Raton London New York

CRC Press is an imprint of the
Taylor & Francis Group, an **informa** business

CRC Press
Taylor & Francis Group
6000 Broken Sound Parkway NW, Suite 300
Boca Raton, FL 33487-2742

Printed on acid-free paper
Version Date: 20140929

International Standard Book Number-13: 978-0-415-52256-4 (Paperback)

Library of Congress Cataloging-in-Publication Data

Zingoni, Alphose.
  Vibration analysis and structural dynamics for civil engineers : essentials and group-theoretic formulations / author, Alphose Zingoni.
      pages cm
  Includes bibliographical references and index.
  ISBN 978-0-415-52256-4 (hardcover : alk. paper) 1. Vibration. I. Title.

TA355.Z54 2015
624.1'76--dc23   2014034109

Contents

List of illustrations

List of tables

Preface

It is not an easy task to write a book that covers the fundamentals of a subject, at the same time going reasonably far enough on a particular aspect of the subject to interest the more specialist reader. This book attempts to do that for the subject of the vibration of engineering systems, later focusing on how symmetry (common in engineering systems) affects vibration behaviour, a topic that, interestingly, has been the subject of research for many years in a number of areas of physics and chemistry, such as crystallography, quantum mechanics and molecular symmetry. After covering the fundamentals (which every student must know), the same approaches that have yielded fruitful insights in studying various problems of symmetry in physics and chemistry are harnessed here for studying vibration problems in engineering.

The book is divided into two parts. The first part covers the essentials of vibration analysis, and is intended for all civil, structural and mechanical engineering students taking a basic course on structural dynamics, as well as practicing engineers wishing to acquaint themselves with the fundamentals of vibration modelling. The treatment is concise yet thorough, with several fully worked out examples. Tutorial questions are featured at the end of the first three chapters to provide further practice for the student. Considerations are extended to continuous systems and the use of the finite-element method in vibration analysis.

The second part introduces a novel approach to the vibration analysis of symmetric systems. The mathematical concepts of group theory are used to simplify a range of vibration problems in structural mechanics, demonstrating the computational benefits of the approach, and also providing valuable insights into the vibration behaviour of symmetric systems as well as logical explanations of certain observed phenomena (e.g. coincident frequencies, similarity of modes and stationary points). This feature sets the book apart from other books on vibration and makes it of interest not only to students and teachers of vibration analysis, but also to researchers involved with the development of computational procedures and the study of new phenomena. A more detailed description follows.

Part I begins (in Chapter 1) with a general introduction to vibration and its modelling. After these preliminaries, treatment then focuses on the vibration of discrete systems or lumped parameter models. Chapter 2 considers the free vibration response of both damped and undamped single degree-of-freedom systems on the basis of a rectilinear spring–mass model, and extends the ideas to a wider range of problems in structural and mechanical engineering through the concept of an equivalent spring stiffness. This is followed by treatment of the forced response to harmonic excitation of one degree-of-freedom systems.

Chapter 3 is concerned with the vibration of systems with two or more degrees of freedom. The equations of motion are formulated on the basis of both the stiffness matrix and the flexibility matrix, and the derivations of these matrices are amply illustrated by reference to both structural and mechanical systems. Formulation of the eigenvalue problem is followed by evaluation of natural frequencies of vibration and mode shapes. The concept of a generalized mass and a generalized stiffness is then introduced, after which the mode superposition method (or modal analysis) is described, and applied to the evaluation of the forced response of multi-degree-of-freedom systems.

The theoretical formulation of the problems of the vibration of continuous systems (distributed parameter modelling) is dealt with in Chapter 4, by reference to the transverse vibration of strings, the axial vibration of rods and the flexural vibration of beams. Closed-form solutions for free vibration response are obtained. The orthogonality of natural modes is demonstrated, leading to the development of modal analysis as a technique for evaluating the dynamic response of a system to a time-varying excitation force.

The essence of the finite-element formulation is presented in Chapter 5. After an outline of the basic theory, stiffness and consistent mass matrices for truss, beam and rectangular plane–stress finite elements are derived. A simple example is used to illustrate the process of assembly of the system eigenvalue problem and solution of the ensuing equations.

Part II begins (in Chapter 6) with an outline of group theory, a description of symmetry groups and a summary of the most important results of associated representation theory. The chapters that follow illustrate how group theory is implemented in simplifying the vibration analysis of various structural systems exhibiting symmetry properties.

By reference to rectilinear shaft–disc torsional and spring–mass extensional systems (Chapter 7), plane structural grids (Chapter 8) and high-tension cable nets (Chapter 9), the computation of natural frequencies and mode shapes for discrete systems is shown to be considerably simplified through the presented group-theoretic formulations. It is seen how group theory predicts all the symmetry types associated with a particular configuration of the structure, where stationary points and nodal lines occur, and why certain frequencies have the same values. These are important insights that help us to better understand the nature of vibration phenomena in structures.

Considerations are then extended to the vibration analysis of continuous systems by numerical methods. A group-theoretic formulation of the finite-difference method is presented in Chapter 10, and applied to the vibration analysis of square and rectangular plates. It is seen how group-theoretic decomposition of the vector space of the problem permits the evaluation of system frequencies within decoupled subspaces, thus avoiding numerical problems associated with the computation of clustered frequencies. In the last chapter of the book (Chapter 11), a group-theoretic formulation for finite elements is developed, and illustrated by reference to solid rectangular elements.

The second part of the book, which is essentially a brief exposition of a relatively new computational approach for special vibration problems, will appeal to all students of vibration analysis keen to learn new methods, irrespective of engineering discipline. This part will also be of considerable interest to researchers concerned with the development of improved computational methods for vibration analysis, and with the understanding of complex vibration phenomena associated with symmetry groups of high order.

Alphose Zingoni
Cape Town, South Africa

Acknowledgments

The author is grateful to his institution, the University of Cape Town, for granting him research and study leave in 2013 for the purpose of completing the writing of this book. Special thanks are due to Angus Rule of the University of Cape Town, who assisted with the preparation of the illustrations. The original book proposal benefitted from the valuable input of Tony Moore of CRC Press (Taylor & Francis), and the helpful suggestions of various anonymous reviewers, for which the author is most thankful. Acknowledgments are also due to Joselyn Banks-Kyle and Tara Nieuwesteeg, both of CRC Press, for professionally steering the book through the production stages. Feedback received from the author's postgraduate students, past and present, is also acknowledged with thanks. Last but not least, the author is indebted to his wife Lydia and children Ratidzo, Tafadzwa and Simbarashe, for their understanding and constant encouragement throughout the writing of the book.

About the author

Alphose Zingoni, PhD, is professor of structural engineering and mechanics in the Department of Civil Engineering at the University of Cape Town. He holds an MSc in structural engineering and a PhD in shell structures, both earned at Imperial College London. He served as dean of the Faculty of Engineering at the University of Zimbabwe from 1996 to 1999, before moving to the University of Cape Town in 1999, where he served as head of the Department of Civil Engineering from 2008 to 2012.

A past recipient of a Royal Commission for the Exhibition of 1851 Postdoctoral Fellowship of the UK (1992–1994), Dr. Zingoni has research interests encompassing shell structures, space structures, vibration analysis and applications of group theory to problems in computational structural mechanics. He has written more than 100 scientific papers on these topics, which have been published in leading international journals and presented at various international conferences worldwide. He has authored two other books: *Shell Structures in Civil and Mechanical Engineering*, published by Thomas Telford (London, 1997), and *Theory and Analysis of Structures*, published by UNESCO (Nairobi, 2000).

In 2001, he founded the Structural Engineering, Mechanics and Computation (SEMC) series of international conferences, now held in Cape Town (South Africa) every 3 years. The proceedings of these conferences, which he edits, feature peer-reviewed research papers on a wide variety of topics, including vibration analysis and structural dynamics, and the related topics of vibration control, seismic response and earthquake-resistant design. Currently Dr. Zingoni serves on the editorial boards of seven international journals. Over the past 10 years, he has served on the scientific committees of up to 50 international conferences in fields related to structural engineering and mechanics.

Throughout his academic career, he has taught various courses at both the undergraduate and postgraduate levels, which have included courses on structural analysis, structural design, plates and shells, vibration analysis and structural dynamics. He has also supervised a large number of post-graduate students undertaking research in these areas.

Dr. Zingoni is a fellow of the Institution of Structural Engineers (London), a fellow of the International Association for Bridge and Structural Engineering (Zurich), a fellow of the South African Academy of Engineering, a member of the Academy of Science of South Africa, a registered professional engineer with the Engineering Council of South Africa and a registered chartered engineer with the Engineering Council of the UK.

Part I

Essentials

Chapter 1

Introduction

1.1 DEFINITIONS, AIMS AND GENERAL CONCEPTS

The study relating the motion of physical systems and the forces causing them is called *dynamics*. The laws of physics governing this motion are the well-known Newton's laws of motion. If the physical system under study is a structure, such as a tall building, an aircraft fuselage and wings, a car chassis, an industrial floor slab or a telecommunications mast, then the subject becomes *structural dynamics*. Of considerable importance is the oscillatory motion of a physical system about a certain equilibrium position. This is called *vibration*. Vibration analysis is the determination of the frequencies of the oscillations with respect to time, the relative positions of the various components of the physical system during its motion, and the amplitudes of the oscillatory motion of these components.

On a general level, dynamic analysis of a structural system is the determination of its *response* to a time-varying load. The term *response* encompasses the displacements, velocities and accelerations of the various components of the system, as well as the strains and stresses induced in the system. However, it is often sufficient to consider the response of a structural system as fully known, once the displacement behaviour has been completely evaluated, since all the other quantities are functions of the displacements. Depending on the nature of the load and the physical characteristics of the system, the response may not necessarily be of a vibratory character.

Structural, mechanical, naval and aeronautical engineers are required to study vibration in order to be able to eliminate it or control it in the design of systems for which excessive vibration is undesirable (for example, bridges, buildings and vehicles), or to enhance it in the design of systems for which vibration is desired (for example, mechanical vibrators for wet concrete in the construction industry, and demolition work, also pertaining to the construction industry).

The physical characteristics of a dynamic system are called its parameters. Knowing these, a suitable mathematical model of the dynamic behaviour of

3

the system is devised, and analysed to determine the response of the system. Two main approaches to physical modelling exist: (1) *discrete parameter* models, in which parameters are assumed to be 'lumped' at selected locations of the system, and (2) *distributed parameter* models, in which parameters are allowed to vary continuously over the system.

In the course of mathematical modelling, the first type of physical model generally results in *ordinary* differential equations in which displacements vary continuously only with respect to time, while the second results in *partial* differential equations in which displacements vary continuously with respect to both time and space. Thus, while distributed parameter models are generally more representative of the actual behaviour of the physical system, they tend to be considerably more difficult to analyse in comparison with discrete parameter models.

1.2 BASIC FEATURES OF A VIBRATING SYSTEM, AND FURTHER CONCEPTS

These may be identified in very broad terms as (1) the *mass m*, which possesses inertia, and can have kinetic energy by virtue of its velocity of motion; (2) the *spring k*, which possesses elasticity, and can store potential energy by virtue of its deformation; (3) the *damper c*, which is assumed to possess neither inertia nor elasticity, and represents the dissipation of energy from the moving system through friction and similar effects; and (4) the *excitation force f(t)*, which is a function of time *t*, and is the agent by which energy is injected into the system, thereby sustaining vibrations. The excitation force may or may not be present.

Figure 1.1 is a diagrammatic representation of the above four features. The mass *m* is connected to a rigid vertical plane through the spring *k* and the damper *c*, which are parallel to each other. Both the excitation force *f(t)* acting upon the mass, and the displacement *x* of the mass relative to its equilibrium position, are in the horizontal direction, as shown. The weight of the mass, being a vertical force, has no effect upon the (horizontal) motion of the mass. The mass is shown supported on rollers to denote that there is no need to consider any horizontal frictional forces between the mass and the

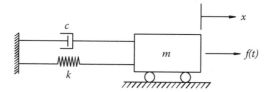

Figure 1.1 Basic features of a vibrating system.

surface upon which it is supported, since all frictional effects in the system, irrespective of their source, are already taken into account by the damper.

The inertial force acting upon the mass m is given by

$$f_m = -m\ddot{x} \tag{1.1}$$

where $\ddot{x}$ is the acceleration of the mass. In the present treatment, a single dot above a variable denotes the first derivative of the variable with respect to time, while two dots denote the second derivative. Thus, $\ddot{x}$ denotes the second derivative of the displacement x with respect to time, which, of course, is the acceleration.

For linear springs, the restoring force in the spring is related to the deformation of the spring (which may be an extension or a compression) through Hooke's law:

$$f_s = kx \tag{1.2}$$

where k is referred to as the *spring constant*, and x, of course, is the displacement of one end of the spring relative to the other. In vibration considerations, x is the displacement coordinate of the oscillating mass relative to its position under conditions of static equilibrium.

The simplest damping model is the *linear viscous dashpot model*, where the damping force is assumed to be proportional to the velocity $\dot{x}$ of the moving mass:

$$f_d = c\dot{x} \tag{1.3}$$

where the parameter c is known as the *coefficient of viscous damping.*

If a system is set in motion by displacing it from its equilibrium position and releasing it, and no excitation force is applied after the initiation of motion, then we say it subsequently undergoes *free vibration*. On the other hand, if the motion of a system is occurring under the influence of a continuing excitation force, then the system is said to undergo *forced vibration*. Depending on whether the damping force in a system is negligible or significant, free or forced vibration may be considered, for analysis purposes, as *undamped* or *damped*.

The number of *degrees of freedom* of a vibrating structural or mechanical system is the number of independent coordinates in space required to fully describe its motion. Discrete parameter systems with one, two and higher numbers of degrees of freedom are illustrated in Figure 1.2.

In the rest of Part I, we will first focus attention on discrete parameter models in Chapter 2 (single degree-of-freedom systems) and Chapter 3 (multi-degree-of-freedom systems), since many structural systems are essentially of this type. We will then turn our attention to the modelling of continuous systems by analytical methods (Chapter 4) and the finite-element method (Chapter 5).

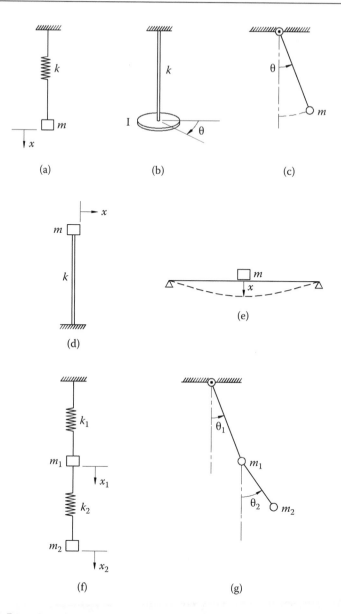

Figure 1.2 Examples of discrete parameter models. One degree-of-freedom models: (a) spring–mass extensional system, (b) shaft–disc torsional system, (c) simple pendulum, (d) beam–mass flexural model of an elevated water tower, (e) simply supported beam carrying a concentrated mass. Two degree-of-freedom models: (f) spring–mass extensional system, (g) double pendulum.

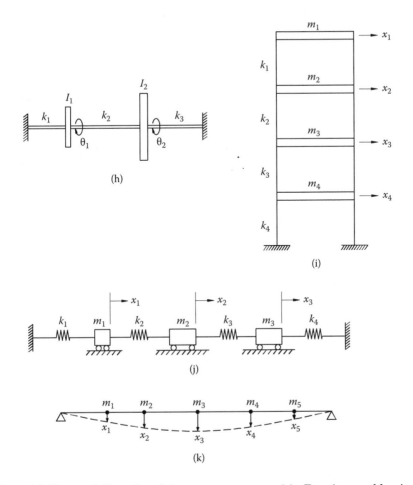

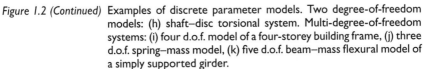

Figure 1.2 (Continued) Examples of discrete parameter models. Two degree-of-freedom models: (h) shaft–disc torsional system. Multi-degree-of-freedom systems: (i) four d.o.f. model of a four-storey building frame, (j) three d.o.f. spring–mass model, (k) five d.o.f. beam–mass flexural model of a simply supported girder.

TUTORIAL QUESTIONS

1.1. What do you understand by the following expressions, within the context of vibration analysis and structural dynamics?
   a. Lumped parameter system
   b. Continuous system
   c. Linear vibration
   d. Nonlinear vibration

    e. Damped system
    f. Undamped system
    g. Free vibration
    h. Forced vibration
    i. Degree of freedom

1.2. How many translational degrees of freedom are possessed by the spring–mass systems shown in Figure Q1.2? In (a) and (b), the systems lie in a plane, with each mass in the latter figure being confined to move in one guided direction only. In (c), which depicts a dynamic model of a train, the spring–mass system lies along a straight line.

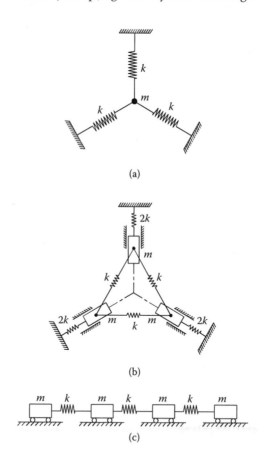

(a)

(b)

(c)

Figure Q1.2 Spring–mass systems of Question 1.2.

Chapter 2

Single degree-of-freedom systems

2.1 BASIC EQUATION OF MOTION

Consider the basic single degree-of-freedom system depicted in Figure 1.1. A free-body diagram of the mass m is shown in Figure 2.1. In general, for a system comprising one or more masses (that is, with one or more degrees of freedom), a free-body diagram of a given mass is a sketch of the mass isolated from the rest of the system, showing all the forces acting upon it.

Summing up all the forces in the direction of motion, we must have, for equilibrium,

$$m\ddot{x} + c\dot{x} + kx = f(t) \tag{2.1}$$

Equation (2.1) is the basic equation of motion for a single degree-of-freedom system. We will first explore the free vibration response of the system, by setting $f(t)$ to zero.

2.2 FREE VIBRATION RESPONSE

The equation of motion for the free vibration of a single degree-of-freedom system is obtained by setting the right-hand side of Equation (2.1) to zero:

$$m\ddot{x} + c\dot{x} + kx = 0 \tag{2.2}$$

In order to solve this homogeneous second-order ordinary differential equation, we assume the displacement of the mass to vary with respect to time in accordance with the following law:

$$x(t) = ae^{\lambda t} \tag{2.3}$$

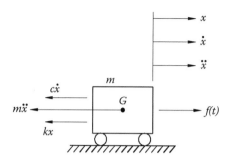

Figure 2.1 Forces on the mass m of a single degree-of-freedom system: kx is the restoring pull of the spring; $c\dot{x}$ is the damping force, in the direction opposite to the direction of motion; $m\ddot{x}$ is the inertial force, acting through the centre of mass G of mass m; and $f(t)$ is the excitation force.

The velocity $\dot{x}(t)$ and acceleration $\ddot{x}(t)$ of the mass follow when the right-hand side of Equation (2.3) is differentiated once and twice, respectively, with respect to time t. Thus,

$$\dot{x}(t) = a\lambda e^{\lambda t} \tag{2.4}$$

$$\ddot{x}(t) = a\lambda^2 e^{\lambda t} \tag{2.5}$$

Substituting expressions (2.3) to (2.5) into the equation of motion (2.2), we obtain

$$ae^{\lambda t}(m\lambda^2 + c\lambda + k) = 0 \tag{2.6}$$

that is,

$$\lambda^2 + \frac{c}{m}\lambda + \frac{k}{m} = 0 \tag{2.7}$$

Equation (2.7) is called the *characteristic equation* of the system. Its roots λ_1 and λ_2 are as follows:

$$\lambda_{1,2} = \frac{1}{2}\left(-\frac{c}{m} \pm \sqrt{\frac{c^2}{m^2} - 4\frac{k}{m}}\right) \tag{2.8}$$

The solution is a linear combination of terms corresponding to the above two roots; from Equation (2.3), this may be written as follows:

$$x(t) = a_1 e^{\lambda_1 t} + a_2 e^{\lambda_2 t} \tag{2.9}$$

Let us introduce the relationships:

$$\frac{k}{m} = \omega_n^2 \tag{2.10}$$

$$\frac{c}{m} = 2\xi\omega_n \tag{2.11}$$

where, for reasons that will become clear in due course, ω_n is called the *undamped natural circular frequency* of the system, and ξ is a parameter called the *viscous damping factor*. Then we may write

$$\lambda_{1,2} = \frac{1}{2}\left(-2\xi\omega_n \pm \sqrt{4\xi^2\omega_n^2 - 4\omega_n^2}\right) = -\xi\omega_n \pm \left(\sqrt{\xi^2 - 1}\right)\omega_n \tag{2.12}$$

Depending on the magnitude of the parameter ξ relative to unity, we have three cases of motion to consider:

1. If $\xi > 1$, the roots in Equation (2.12) are real, distinct and negative. Both terms of the solution for $x(t)$ – refer to Equation (2.9) – diminish exponentially with time t, and there is no oscillatory motion. Unless otherwise implied, the term *oscillatory* or *vibratory* implies a motion whereby the mass crosses its equilibrium position at least once (that is, the displacement changes from positive to negative, or vice versa, at least once during the motion of the mass). If we let

$$\omega_d = \omega_n\sqrt{\left|\xi^2 - 1\right|} = \omega_n\sqrt{\xi^2 - 1} \tag{2.13}$$

where ω_d is called the *damped natural circular frequency* of the system (for reasons that will become clear very shortly), then we may write the roots in Equation (2.12) as follows:

$$\lambda_{1,2} = -\xi\omega_n \pm \omega_d \tag{2.14}$$

Equation (2.9) then becomes

$$x(t) = a_1 e^{(-\xi\omega_n + \omega_d)t} + a_2 e^{(-\xi\omega_n - \omega_d)t} = e^{-\xi\omega_n t}\left(a_1 e^{\omega_d t} + a_2 e^{-\omega_d t}\right) \tag{2.15}$$

The term in the brackets of the final expression may be written as a linear combination of the hyperbolic sine and hyperbolic cosine of $\omega_d t$, so that the expression for $x(t)$ becomes

$$x(t) = e^{-\xi\omega_n t}\left(A_1 \sinh\omega_d t + A_2 \cosh\omega_d t\right) \tag{2.16}$$

The constants A_1 and A_2 are determined from the initial conditions of the problem.

2. If $\xi = 1$, then, from relation (2.11),

$$c = 2m\omega_n = c_{cr} \qquad (2.17)$$

where c_{cr} is called the *critical damping coefficient*. From relations (2.11) and (2.17), it follows that

$$\xi = \frac{c}{2m\omega_n} = \frac{c}{c_{cr}} \qquad (2.18)$$

In this case, both the roots of λ are equal to $-\omega_n$, so that, from the theory of homogeneous ordinary linear differential equations (refer to any text on the subject), the solution for $x(t)$ now takes the form

$$x(t) = (a_1 + a_2 t)\, e^{\lambda t} = (a_1 + a_2 t)\, e^{-\omega_n t} \qquad (2.19)$$

Again, there are no oscillations. For the value $\xi = 1$, the system is said to be *critically damped*, and c has the value c_{cr}. From a physical standpoint, c_{cr} is the smallest value of the damping coefficient for which no oscillations occur.

3. If $\xi < 1$, then the roots in Equation (2.12) are complex, and we may write

$$\lambda_{1,2} = -\xi\omega_n \pm i\left(\sqrt{1-\xi^2}\right)\omega_n = -\xi\omega_n \pm i\omega_d \qquad (2.20)$$

so that

$$x(t) = a_1 e^{(-\xi\omega_n + i\omega_d)t} + a_2 e^{(-\xi\omega_n - i\omega_d)t} = e^{-\xi\omega_n t}\left(a_1 e^{i\omega_d t} + a_2 e^{-i\omega_d t}\right) \qquad (2.21)$$

The term in the brackets of the final expression may be written as a linear combination of the trigonometric cosine and trigonometric sine of $\omega_d t$, which in turn may be written as a single trigonometric term, so that the expression for $x(t)$ becomes

$$x(t) = e^{-\xi\omega_n t}\left(A_1 \cos\omega_d t + A_2 \sin\omega_d t\right) = Ae^{-\xi\omega_n t}\cos\left(\omega_d t - \alpha\right) \qquad (2.22)$$

showing that the motion is an oscillatory *simple harmonic motion of circular frequency* ω_d (which is why we called ω_d the *damped natural circular frequency* of the system) and *amplitude* $Ae^{-\xi\omega_n t}$. We note that the amplitude decreases exponentially with time. In the final expression, A and α are constants to be determined from the initial conditions of the problem. The constant α is called the *phase angle* of the motion.

For the case $\xi < 1$, the system is said to be *underdamped*.

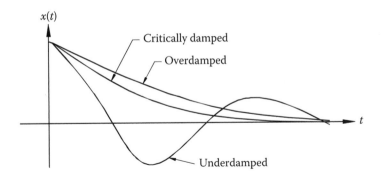

Figure 2.2 Displacement variation with time for overdamping, critical damping and under-damping, for a single degree-of-freedom system undergoing free vibration.

The three cases of overdamping, critical damping and underdamping are depicted in Figure 2.2. Note that the critically damped case represents the level of damping for which the system will return to the equilibrium position in the shortest possible time, without undergoing any oscillation(s). Clearly, critical damping has an important significance in the design of structural systems against vibration.

Let us examine the case of underdamping in more detail, since it is the only case associated with oscillation or vibration, in the sense defined earlier.

The initial conditions (that is, the displacement and velocity of the mass at time $t = 0$) may be expressed as follows:

$$x(0) = x_0 \tag{2.23}$$

$$\dot{x}(0) = \dot{x}_0 \tag{2.24}$$

where x_0 and $\dot{x}_0$ are assumed to be known. For instance, if the mass m is set in free vibration by giving it an initial displacement of magnitude 0.1 metres from its equilibrium position, and then releasing it from rest, then $x_0 = 0.1\text{m}$ and $\dot{x}_0 = 0$.

Before we can impose both initial conditions upon the solution for $x(t)$, we require the expression for the velocity $\dot{x}(t)$. Differentiating expression (2.22) once with respect to t, we obtain

$$\dot{x}(t) = A\left[-\xi\omega_n e^{-\xi\omega_n t}\cos(\omega_d t - \alpha) - \omega_d e^{-\xi\omega_n t}\sin(\omega_d t - \alpha)\right]$$

$$= -Ae^{-\xi\omega_n t}\left[\xi\omega_n \cos(\omega_d t - \alpha) + \omega_d \sin(\omega_d t - \alpha)\right] \tag{2.25}$$

Substituting the initial conditions into expressions (2.22) and (2.25) for $x(t)$ and $\dot{x}(t)$ respectively, we obtain

$$x_0 = A\cos\alpha \tag{2.26}$$

$$\dot{x}_0 = -A\left(\xi\omega_n \cos\alpha - \omega_d \sin\alpha\right) \tag{2.27}$$

From the first of the above equations, we obtain the relationship

$$A = \frac{x_0}{\cos\alpha} \tag{2.28}$$

Using this result in Equation (2.27), we may write

$$\dot{x}_0 = -x_0\left(\xi\omega_n - \omega_d \tan\alpha\right) \tag{2.29}$$

Rearranging, we obtain

$$\tan\alpha = \frac{\xi\omega_n + \dfrac{\dot{x}_0}{x_0}}{\omega_d} = \frac{\xi\omega_n x_0 + \dot{x}_0}{\omega_d x_0} \tag{2.30}$$

so that the value of the angular constant α is given by

$$\alpha = \tan^{-1}\left(\frac{\xi\omega_n x_0 + \dot{x}_0}{\omega_d x_0}\right) \tag{2.31}$$

Figure 2.3 shows a right-angled triangle satisfying relation (2.30). The length of the hypotenuse is given by Pythagoras' theorem. From the sketch, we have

$$\cos\alpha = \frac{\omega_d x_0}{\sqrt{\left(\xi\omega_n x_0 + \dot{x}_0\right)^2 + \omega_d^2 x_0^2}} \tag{2.32}$$

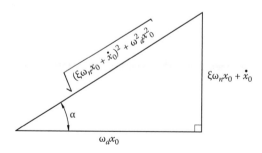

Figure 2.3 Trigonometric representation of Equation (2.30).

so that the value of the constant A may finally be written as

$$A = \frac{1}{\omega_d} \sqrt{\left(\xi\omega_n x_0 + \dot{x}_0\right)^2 + \omega_d^2 x_0^2}$$

(2.33)

Substituting the now known values of the constants A and α into expression (2.22), we finally obtain

$$x(t) = \left\{ \frac{1}{\omega_d} \sqrt{\left(\xi\omega_n x_0 + \dot{x}_0\right)^2 + \omega_d^2 x_0^2} \right\} e^{-\xi\omega_n t} \cos\left[\omega_d t - \tan^{-1}\left(\frac{\xi\omega_n x_0 + \dot{x}_0}{\omega_d x_0} \right) \right]$$

(2.34)

The precise damping characteristics of a structural or mechanical system are usually quite difficult to quantify. However, we may obtain a good estimate of the damping behaviour by determining an *equivalent viscous damping ratio* ξ of the system, through experimental measurements.

Consider Figure 2.4, which is a sketch of the displacement variation with respect to time for an underdamped system. From Equation (2.22), the ratio A_r / A_{r+1} of the amplitudes of any two successive oscillations is given by

$$\frac{A_r}{A_{r+1}} = \frac{Ae^{-\xi\omega_n t_r}}{Ae^{-\xi\omega_n t_{r+1}}} = e^{\xi\omega_n(t_{r+1}-t_r)}$$

(2.35)

The interval $(t_{r+1} - t_r)$ is, by definition, the period T of the oscillations (refer to Figure 2.4). Now, for any simple harmonic motion of circular frequency ω, the period is given by

$$T = \frac{2\pi}{\omega}$$

(2.36)

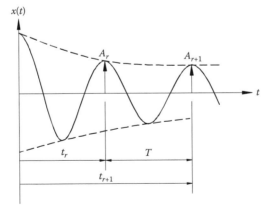

Figure 2.4 Displacement variation with time for an underdamped single degree-of-freedom system undergoing free vibration. The plot shows the relative amplitudes of two successive oscillations.

Since the circular frequency of the present problem is ω_d (refer to Equation (2.22)), we can replace $(t_{r+1} - t_r)$ in Equation (2.35) by $2\pi/\omega_d$, so that

$$\frac{A_r}{A_{r+1}} = e^{2\pi\xi(\omega_n/\omega_d)} \qquad (2.37)$$

Using relation (2.13) to eliminate ω_d, we obtain

$$\frac{A_r}{A_{r+1}} = e^{2\pi\xi/\sqrt{\left|\xi^2-1\right|}} \approx e^{2\pi\xi} \qquad (2.38)$$

assuming $\xi^2 \ll 1$ (that is, ξ^2 is much less than 1, implying that the damping is relatively weak).

Now, the series expansion of $e^{2\pi\xi}$ is

$$e^{2\pi\xi} = 1 + (2\pi\xi) + \frac{(2\pi\xi)^2}{2!} + \frac{(2\pi\xi)^3}{3!} + \dots$$

$$\approx 1 + 2\pi\xi \qquad (2.39)$$

the neglecting of all terms after the second being justified since these would all be very small in comparison with the first two terms. This approximation is consistent with our earlier assumption that $\xi^2 \ll 1$, but notice the term $2\pi\xi$ is not being neglected in relation to 1, since ξ itself, while relatively small, need not be negligible in comparison with 1.

Expression (2.38) may therefore be written as follows:

$$\frac{A_r}{A_{r+1}} \approx 1 + 2\pi\xi \qquad (2.40)$$

Rearranging, we obtain

$$\xi \approx \frac{A_r - A_{r+1}}{2\pi A_{r+1}} \qquad (2.41)$$

Thus, by accurately measuring the amplitudes A_r and A_{r+1} of any two successive oscillations, and using these values in Equation (2.41), we can arrive at a reasonable estimate of the equivalent viscous damping ratio ξ for any lightly to moderately damped system within the underdamped regime of behaviour.

Let us consider the special case when there is absolutely no damping. In this case, we set the parameter c equal to zero, so that the equation of motion for free vibration (that is, expression (2.2)) reduces to

$$m\ddot{x} + kx = 0 \qquad (2.42)$$

We assume a solution of the same form given by Equation (2.3), and substitute expressions (2.3) and (2.5) for x and $\ddot{x}$, respectively, into the above equation. This gives us the characteristic equation

$$\lambda^2 + \frac{k}{m} = 0 \tag{2.43}$$

which, upon making use of relation (2.10), may be rewritten as

$$\lambda^2 + \omega_n^2 = 0 \tag{2.44}$$

The roots of Equation (2.44) are as follows:

$$\lambda = \pm i\omega_n \tag{2.45}$$

The solution for $x(t)$ is therefore

$$x(t) = a_1 e^{i\omega_n t} + a_2 e^{-i\omega_n t} \tag{2.46}$$

which may be rewritten as

$$x(t) = A_1 \sin\omega_n t + A_2 \cos\omega_n t \tag{2.47}$$

where A_1 and A_2 are constants to be determined from the initial conditions. For example, if the initial conditions are as given by Equations (2.23) and (2.24), then

$$A_2 = x_0 \tag{2.48}$$

$$A_1 = \frac{\dot{x}_0}{\omega_n} \tag{2.49}$$

so that

$$x(t) = \left(\frac{\dot{x}_0}{\omega_n}\right)\sin\omega_n t + x_0 \cos\omega_n t \tag{2.50}$$

For a system that is displaced from its equilibrium position by an amount x_0, and simply released without imparting on it an initial velocity, we have $\dot{x}_0 = 0$, so that

$$x(t) = x_0 \cos\omega_n t \tag{2.51}$$

This is simple harmonic motion of amplitude x_0 and circular frequency ω_n. The natural period of the motion is, of course, given by

$$T_n = \frac{2\pi}{\omega_n} \tag{2.52}$$

and the natural frequency (number of oscillations per second) follows as the reciprocal of the natural period:

$$f_n = \frac{1}{T_n} = \frac{\omega_n}{2\pi} \qquad (2.53)$$

This motion and its associated initial conditions are depicted in Figure 2.5(a).

In the general case of both x_0 and $\dot{x}_0$ being nonzero, we may rewrite expression (2.47) in the form

$$x(t) = A\cos(\omega_n t - \alpha) \qquad (2.54)$$

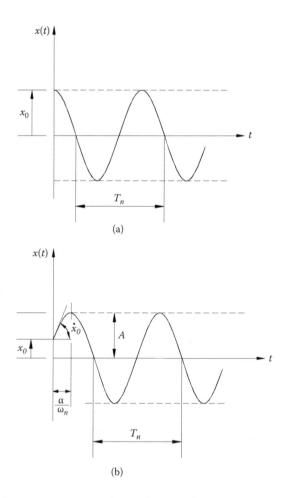

Figure 2.5 Displacement variation with time for an undamped single degree-of-freedom system undergoing free vibration: (a) initial conditions $x(0) = x_0 \neq 0$ and $\dot{x}(0) = \dot{x}_0 = 0$, (b) initial conditions $x(0) = x_0 \neq 0$ and $\dot{x}(0) = \dot{x}_0 \neq 0$.

where the constants A and α are the amplitude and phase angle of the motion. This motion is simple harmonic motion, the circular frequency being ω_n (which is why we earlier on called ω_n the *undamped natural circular frequency* of the system). Substituting the generalized initial conditions into Equation (2.54), we find that

$$A = \sqrt{\left(\frac{\dot{x}_0}{\omega_n}\right)^2 + x_0^2} \tag{2.55}$$

$$\alpha = \tan^{-1}\left(\frac{(\dot{x}_0 / \omega_n)}{x_0}\right) \tag{2.56}$$

so that

$$x(t) = \left\{\sqrt{\left(\frac{\dot{x}_0}{\omega_n}\right)^2 + x_0^2}\right\}\cos\left[\omega_n t - \tan^{-1}\left(\frac{(\dot{x}_0 / \omega_n)}{x_0}\right)\right] \tag{2.57}$$

This motion and its associated initial conditions are depicted in Figure 2.5(b).

2.3 EQUIVALENT SPRING STIFFNESSES FOR VARIOUS STRUCTURAL AND MECHANICAL SYSTEMS

Figure 2.6 shows a vertical circular shaft of uniform diameter d and length l, rigidly built-in at the upper end and rigidly attached to a heavy circular disc at the lower end, such that the shaft and the disc share a common

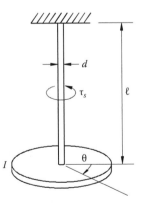

Figure 2.6 Vertical circular shaft of diameter d and length l, rigidly clamped at the upper end and carrying at the lower end a circular disc of moment of inertia I.

vertical axis of rotational symmetry. The moment of inertia of the shaft about this common axis is negligible, but that of the disc is significant, being of magnitude I. The motion of this system is rotational; it is described by the coordinate θ, which is the angular displacement of any radius of the disc measured relative to the equilibrium position of that radius.

If the shaft is given a small rotation of magnitude θ at its lower end and held in that position, the torque tending to restore the disc to its equilibrium position is given by

$$\tau_s = \left(\frac{GJ}{l}\right)\theta \tag{2.58}$$

where

$$J = \frac{\pi d^4}{32} \tag{2.59}$$

is the second moment of area of any (horizontal) cross section of the circular shaft about the (vertical) perpendicular axis through its centre, l is the length of the shaft and G is the shear modulus of rigidity of the material of the shaft (refer to any text on the subject of strength of materials).

As a result of the angular acceleration the disc has during the torsional oscillations of the system, an inertial torque acts upon the disc. This is given by

$$\tau_m = -I\ddot{\theta} \tag{2.60}$$

where $\ddot{\theta}$ is the second derivative of θ with respect to time t (that is, the angular acceleration of the disc).

Assuming undamped free vibrations, and summing up all torques acting upon the disc, we have, for equilibrium,

$$I\ddot{\theta} + \left(\frac{GJ}{l}\right)\theta = 0 \tag{2.61}$$

that is,

$$I\ddot{\theta} + k_t\theta = 0 \tag{2.62}$$

which is the equation of motion of the disc. Thus, for torsional oscillations, moment of inertia I of the disc takes the place of mass m, while the torsional stiffness k_t of the shaft ($= GJ/l$) takes the place of spring constant k.

For each of the systems shown in Figure 2.7, we use the interpretation, based upon the basic Hookean relationship $P = k\delta$ linking spring force P with spring extension δ through the spring constant k, that if we make δ equal to 1, then the force required to maintain such a configuration of

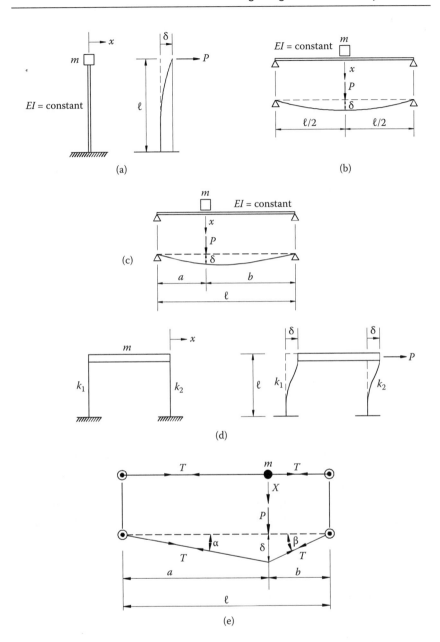

Figure 2.7 Determination of equivalent spring stiffnesses for various single degree-of-freedom systems: (a) cantilever, (b, c) simply supported beams, (d) single-storey building frame, (e) taut string or cable.

unit deflection is numerically equal to k. This relationship assumes, of course, that displacements are small in comparison with the dimensions of the structure undergoing deformation, so that the response of the structure is essentially linear (that is, deformations are directly proportional to the magnitude of the forces causing them).

The system shown in Figure 2.7(a) comprises a vertical cantilever column of constant flexural rigidity EI and negligible mass, carrying a concentrated mass m at its free end. The relevant vibration behaviour here is the lateral (that is, horizontal) motion x of the mass m, with the column serving as the spring for the system. To obtain the equivalent spring stiffness k for the system, we note that if a horizontal force P is applied at the top of the cantilever, the horizontal deflection occurring there is given by the well-known expression

$$\delta = \frac{Pl^3}{3EI} \tag{2.63}$$

so that

$$P = \left(\frac{3EI}{l^3}\right)\delta \tag{2.64}$$

Thus, for the system shown in Figure 2.7(a),

$$k = \frac{3EI}{l^3} \tag{2.65}$$

In Figure 2.7(b) is shown a simply supported beam of constant flexural rigidity EI and length l, carrying a concentrated mass m at midspan. Assuming the beam itself is of negligible mass, we observe that the system has one degree of freedom, that is, the vertical motion x of the mass m. The vertical deflection under a vertical point load P applied at the midspan location of a simply supported beam is given by the well-known formula

$$\delta = \frac{Pl^3}{48EI} \tag{2.66}$$

Rearranging Equation (2.66), we obtain

$$P = \left(\frac{48EI}{l^3}\right)\delta \tag{2.67}$$

so that

$$k = \frac{48EI}{l^3} \tag{2.68}$$

If the concentrated mass m is carried at an arbitrary distance a from the left-hand end of the simply supported beam, as shown in Figure 2.7(c), we require to know the vertical deflection under a vertical point load P positioned at a distance a from the left-hand end of the beam. This may readily be calculated by any of the standard methods for determining deflections in statically determinate beams, and is found to be

$$\delta = \frac{Pa^2b^2}{3EIl} \tag{2.69}$$

where, from Figure 2.7(c), $b = l - a$.

We rearrange this expression to obtain

$$P = \left(\frac{3EIl}{a^2b^2}\right)\delta \tag{2.70}$$

from which we conclude

$$k = \frac{3EIl}{a^2b^2} \tag{2.71}$$

In the single-storey building frame shown in Figure 2.7(d), the mass m of the system is assumed to be concentrated at roof slab level. The structure responds to lateral load by swaying sideways, so that the degree of freedom x of the system is a horizontal displacement. The upper ends of the vertical columns are rigidly connected to the roof slab. As the rigidity of this slab against rotation in the vertical plane is much larger than that of the vertical columns, we assume the vertical columns are effectively restrained against rotation at their upper end, and they do not undergo any rotation there. In the present example, the lower ends of the columns are assumed to be rigidly built into the foundation, so that these, like the upper ends, also do not undergo any rotation. Now the transverse force P required at each end of a beam element, to maintain a deformation configuration of the beam element whereby one end has a transverse displacement of magnitude δ relative to the other end while both ends are not allowed to rotate, is known from basic structural analysis to be

$$P = \left(\frac{12EI}{l^3}\right)\delta \tag{2.72}$$

implying that

$$k = \frac{12EI}{l^3} \tag{2.73}$$

The arrangement shown in Figure 2.7(d) is sometimes referred to as a shear frame. If a lateral force P is applied to the top of the frame, both the left and the right sides of the rigid slab sway by an equal amount δ, so that

$$P = \left(\frac{12E_1I_1}{l^3}\right)\delta + \left(\frac{12E_2I_2}{l^3}\right)\delta = k_1\delta + k_2\delta = (k_1 + k_2)\delta \tag{2.74}$$

where

$$k_1 = \frac{12E_1I_1}{l^3}; \quad k_2 = \frac{12E_2I_2}{l^3} \tag{2.75}$$

refer to the net stiffness of columns on the left and the right, respectively, assuming the columns all have the same length.

Thus, for the system,

$$k = k_1 + k_2 = \frac{12E_1I_1}{l^3} + \frac{12E_2I_2}{l^3} \tag{2.76}$$

Finally, let us consider an essentially weightless string, stretched taut under a high tension force T. A concentrated mass m is attached to the string at a distance a from the left-hand end, as shown in Figure 2.7(e). Small transverse vibrations of the mass m, defined by the vertical displacement coordinate x measured relative to the equilibrium position of the mass under static conditions, do not appreciably alter the magnitude of the prestress tension T in the string. If we apply a vertical force P at a distance a from the left-hand end of the string, a deflection δ will occur under the point load P, as shown in Figure 2.7(e). For vertical equilibrium at the loaded point on the string, we must have

$$P = T\sin\alpha + T\sin\beta \approx T\left(\frac{\delta}{a} + \frac{\delta}{b}\right) = T\left(\frac{a+b}{ab}\right)\delta \tag{2.77}$$

Therefore,

$$k = \left(\frac{a+b}{ab}\right)T \tag{2.78}$$

2.4 RESPONSE TO HARMONIC EXCITATION

Whereas all along the system, once set in motion, has been assumed to vibrate freely under the influence of no external forces, we will now consider its response if a periodic external loading acts upon the system. We chose to concentrate on harmonic-type excitations, since many time-varying loadings encountered in real situations can be represented as Fourier series expansions involving harmonic terms. Let the loading be a harmonic excitation of amplitude G and circular frequency Ω, so that Equation (2.1) becomes

$$m\ddot{x} + c\dot{x} + kx = G\sin\Omega t \tag{2.79}$$

We start off by initially assuming damping is negligible, and set c equal to zero in the above equation, thus obtaining

$$m\ddot{x} + kx = G\sin\Omega t \tag{2.80}$$

The general solution of the above nonhomogeneous differential equation consists of a homogeneous component x_H and a particular solution x_P. That is,

$$x(t) = x_H + x_P \tag{2.81}$$

The homogeneous component, being the solution to Equation (2.80) with the right-hand side set equal to zero, is, of course, the undamped free vibration response of the system, which has already been dealt with in earlier sections. We will therefore focus attention on the determination of the particular solution.

We assume the particular solution to be of the same form as the excitation force, that is,

$$x_P(t) = A\sin\Omega t \tag{2.82}$$

where A is a constant amplitude.

The velocity and acceleration then follow as

$$\dot{x}_P(t) = A\Omega\cos\Omega t \tag{2.83}$$

$$\ddot{x}_P(t) = -A\Omega^2\sin\Omega t \tag{2.84}$$

Substituting expressions (2.82) and (2.84) into the equation of motion (Equation (2.80)), we obtain

$$-mA\Omega^2\sin\Omega t + kA\sin\Omega t = G\sin\Omega t \tag{2.85}$$

which, upon dividing throughout by $\sin\Omega t$, simplifies to

$$A(k - m\Omega^2) = G \tag{2.86}$$

giving

$$A = \frac{G}{k - m\Omega^2} = \frac{G/k}{1 - \frac{m}{k}\Omega^2} = \frac{G/k}{1 - \frac{\Omega^2}{\omega_n^2}} = \frac{G/k}{1 - r^2} \quad (2.87)$$

recalling that k/m is equal to ω_n^2 (Equation (2.10)); the parameter r is called the *frequency ratio*, being the ratio of the circular frequency of the excitation to the undamped natural circular frequency of the system:

$$r = \frac{\Omega}{\omega_n} \quad (2.88)$$

The numerator term G/k in Equation (2.87) is a measure of the static deflection due to the peak value of the excitation force.

With the constant amplitude A now known, the particular solution becomes

$$x_P(t) = \left(\frac{G/k}{1-r^2}\right)\sin\Omega t \quad (2.89)$$

Combining the above particular solution with the homogeneous solution as given by Equation (2.47), we obtain the general solution as

$$x(t) = x_P(t) + x_H(t)$$

$$= \left(\frac{G/k}{1-r^2}\right)\sin\Omega t + A_1\sin\omega_n t + A_2\cos\omega_n t \quad (2.90)$$

where A_1 and A_2 are constants to be determined from the initial conditions of the problem. If these initial conditions are as given by Equations (2.23) and (2.24), we obtain

$$A_2 = x_0 \quad (2.91)$$

$$A_1 = \frac{\dot{x}_0}{\omega_n} - \left(\frac{G/k}{1-r^2}\right)\frac{\Omega}{\omega_n} = \frac{\dot{x}_0}{\omega_n} - \frac{G}{k}\left(\frac{r}{1-r^2}\right) \quad (2.92)$$

When damping is not negligible (that is, $c \neq 0$), the *steady-state response* of the system will be *out of phase* with the excitation. We therefore assume a particular solution of the form

$$x_P(t) = A\sin(\Omega t - \alpha) \quad (2.93)$$

where, as before, A is a constant amplitude; the constant α is the phase angle by which the system response lags behind the excitation.

From Equation (2.93), we obtain the derivatives

$$\dot{x}_P(t) = A\Omega\cos(\Omega t - \alpha) \tag{2.94}$$

$$\ddot{x}_P(t) = -A\Omega^2\sin(\Omega t - \alpha) \tag{2.95}$$

Substituting expressions (2.93) to (2.95) into the equation of motion (Equation (2.79)), we have

$$-mA\Omega^2\sin(\Omega t - \alpha) + cA\Omega\cos(\Omega t - \alpha) + kA\sin(\Omega t - \alpha) = G\sin\Omega t \tag{2.96}$$

which, upon dividing throughout by A and rearranging, becomes

$$\left(k - m\Omega^2\right)\sin(\Omega t - \alpha) + c\Omega\cos(\Omega t - \alpha) = \frac{G}{A}\sin\Omega t \tag{2.97}$$

This relationship is represented graphically in Figure 2.8, where vectors of length $(k - m\Omega^2)$, $c\Omega$ and G/A have been plotted in the real-imaginary plane, with orientations to the reference axes that satisfy Equation (2.97).

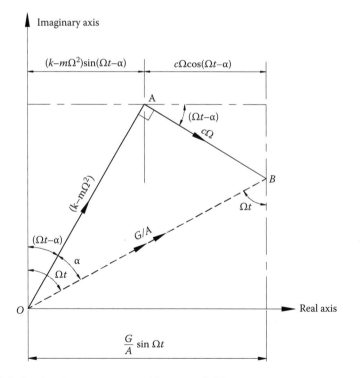

Figure 2.8 Graphical representation of Equation (2.97).

The diagram is self-explanatory. From the triangle OAB, which is right-angled, we have, by Pythagoras' theorem

$$\left(\frac{G}{A}\right)^2 = \left(k - m\Omega^2\right)^2 + \left(c\Omega\right)^2 \tag{2.98}$$

from which we obtain

$$A = \frac{G}{\sqrt{\left(k - m\Omega^2\right)^2 + \left(c\Omega\right)^2}}$$

$$= \frac{G/k}{\sqrt{\left(1 - \frac{m}{k}\Omega^2\right)^2 + \left(\frac{c}{k}\Omega\right)^2}}$$

$$= \frac{G/k}{\sqrt{\left(1 - \frac{\Omega^2}{\omega_n^2}\right)^2 + \left(\frac{2\xi}{\omega_n}\Omega\right)^2}}$$

$$= \frac{G/k}{\sqrt{\left(1 - r^2\right)^2 + \left(2\xi r\right)^2}} \tag{2.99}$$

recalling, from relations (2.10) and (2.11), that $k/m = \omega_n^2$ and $c/m = 2\xi\omega_n$ (so that $c/k = 2\xi/\omega_n$), and also making use of relation (2.88).

From the same triangle OAB, we may also write

$$\tan\alpha = \frac{c\Omega}{k - m\Omega^2} = \frac{\dfrac{c}{k}\Omega}{1 - \dfrac{m}{k}\Omega^2} = \frac{\dfrac{2\xi}{\omega_n}\Omega}{1 - \dfrac{\Omega^2}{\omega_n^2}} = \frac{2\xi r}{1 - r^2} \tag{2.100}$$

The parameter

$$M = \frac{1}{\sqrt{\left(1 - r^2\right)^2 + \left(2\xi r\right)^2}} = \frac{A}{G/k} \tag{2.101}$$

is called the *magnification factor*. In physical terms, M is the ratio of the amplitude A of the steady-state response to the static displacement G/k due to the peak value of the excitation force.

Figure 2.9 shows the magnification factor M and phase angle α plotted against the frequency ratio r.

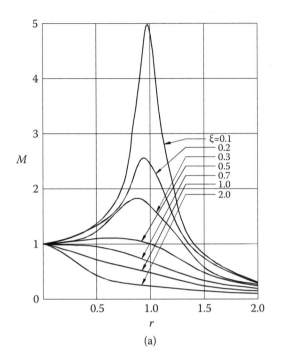

(a)

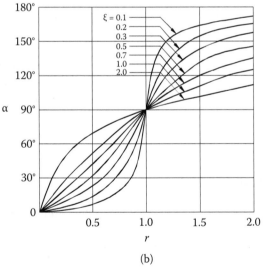

(b)

Figure 2.9 Response of single degree-of-freedom system to harmonic excitation, for various levels of damping: (a) magnification factor *M* versus frequency ratio *r*, (b) phase angle α versus frequency ratio *r*.

A number of important observations may be made regarding the properties of the steady-state response (that is, the system response corresponding to the particular solution).

We note that the steady-state response is harmonic (Equation (2.93)) and of the same frequency as the exciting force. The amplitude of the steady-state response is a function of the amplitude G and frequency Ω of the exciting force, as well as of the natural frequency ω_n and damping factor ξ of the system.

As shown in Figure 2.9(a), the magnification factor M may vary from anything just above zero to values well in excess of 1. *Resonance* is the condition when the frequencies of the exciting force and the system coincide, that is, $\Omega = \omega_n$, so that $r = 1$. At resonance, the amplitude of the steady-state response is limited only by the level of damping in the system, since $M = 1/(2\xi)$. Under resonance conditions, $\tan \alpha = \infty$ so that $\alpha = 90°$, showing that the response lags behind the excitation by 90°.

In general, the phase angle α ($0 \le \alpha \le 180°$) varies with the parameters r and ξ, being either 0 or 180° if $\xi = 0$ (no damping), and exactly 90° if $r = 1$ (at resonance).

Superimposing the *steady-state response* (particular solution) with the *transient response* (homogeneous solution; the term *transient* denotes its dying-out nature), we obtain the total response as follows:

$$x(t) = \frac{G/k}{\sqrt{\left(1-r^2\right)^2 + \left(2\xi r\right)^2}} \sin\left(\Omega t - \alpha_P\right) + A e^{-\xi\omega_n t} \cos\left(\omega_d t - \alpha_H\right) \quad (2.102)$$

where the constants A and α_H are to be determined from the initial conditions. In writing out this final expression, it has been necessary to distinguish the α of the particular solution (which is already known; see Equation (2.100)) from the α of the homogeneous solution (refer to Equation (2.22)), by the subscripted symbols α_P and α_H, respectively.

TUTORIAL QUESTIONS

Unless otherwise stated, assume damping is negligible in all the questions.

2.1. Figure Q2.1 shows a mass m attached to the end of an inextensible string of length l, the other end of the string being attached to a fixed point C, such that the whole forms a pendulum swinging about point C. Derive an expression for the natural frequency of the pendulum, for small values of the angular displacement θ.

2.2. A concentrated mass m is attached to the free end of a horizontal cantilever beam of length l and constant flexural rigidity EI,

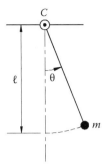

Figure Q2.1 Pendulum of Question 2.1.

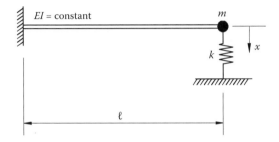

Figure Q2.2 Horizontal cantilever beam of Question 2.2.

as shown in Figure Q2.2. The mass m is also attached to the upper
end of a vertical spring of stiffness k, the lower end of the spring
being connected to a rigid horizontal plane. Derive an expression for the natural frequency of small vertical oscillations of the
mass m, assuming the mass of the beam is negligible.

2.3. A simply supported beam of length l and constant flexural rigidity EI is provided with an elastic vertical support at midspan.
The mass m of the beam is assumed to be lumped at midspan.
A vertical spring of stiffness k, with its upper end attached to the
mass m and its lower end connected to a rigid horizontal plane,
simulates the midspan elastic support, as depicted in Figure Q2.3.
Assuming the amplitude of motion of the mass m is small, what is
the period of the oscillations?

2.4. A heavy horizontal beam of mass m is monolithically connected
to three vertical columns of heights and flexural rigidities as indicated in Figure Q2.4. The lower ends of the columns are built
into a rigid foundation, while the horizontal beam is also suffi-
ciently rigid to prevent rotation of the columns at their top ends.

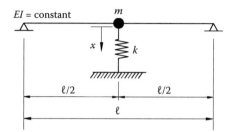

Figure Q2.3 Simply supported beam of Question 2.3.

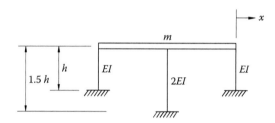

Figure Q2.4 Beam–column system of Question 2.4.

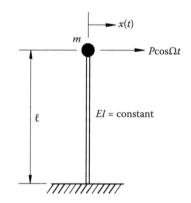

Figure Q2.5 Beam–mass cantilever system of Question 2.5.

Assuming the masses of the columns are negligible in comparison with that of the beam, obtain the period for small lateral vibrations of the system.

2.5. Obtain the response $x(t)$ of the beam–mass cantilever system shown in Figure Q2.5, when the mass is subjected to a harmonic excitation $P\cos\Omega t$. The initial conditions are $x(0) = e$, $\dot{x}(0) = v$.

Chapter 3

Systems with more than one degree of freedom

3.1 INTRODUCTORY REMARK

For a system with n ($n > 1$) degrees of freedom, there are n *natural frequencies*. Some of these frequencies may be equal to each other, particularly where the system has one or more symmetry properties (this subject will be dealt with in Part II). Associated with each natural frequency is a *mode* of vibration, which describes the configuration or shape of the system as it vibrates.

3.2 EQUATIONS OF MOTION

We will illustrate the derivation of these by reference to a three degree-of-freedom spring–mass system with linear viscous damping, the general arrangement of this system being as shown in Figure 3.1(a).

The masses m_1, m_2 and m_3 have degrees of freedom x_1, x_2 and x_3, respectively, and are subjected to excitation forces $f_1(t)$, $f_2(t)$ and $f_3(t)$, respectively, as shown in Figure 3.1(a). Masses m_1 and m_2 are connected through a spring of *spring constant k_2* and a damper of *coefficient of viscous damping c_2*, arranged in parallel as shown. Similarly, masses m_2 and m_3 are connected through the spring-and-damper pair $\{k_3, c_3\}$. The left-hand end of mass m_1 is connected to a rigid vertical plane through the spring-and-damper pair $\{k_1, c_1\}$; the right-hand end of mass m_3 is connected to a rigid vertical plane through the spring-and-damper pair $\{k_4, c_4\}$.

Figure 3.1(b) shows the free-body diagrams of masses m_1, m_2 and m_3, when these masses are assumed to be (1) displaced towards the right from their respective equilibrium positions by amounts x_1, x_2 and x_3, respectively, (2) moving towards the right with velocities of $\dot{x}_1$, $\dot{x}_2$ and $\dot{x}_3$, respectively, and (3) accelerating towards the right with accelerations of $\ddot{x}_1$, $\ddot{x}_2$ and $\ddot{x}_3$, respectively. Notice that the restoring force in any spring is equal to the spring constant of that spring multiplied by the displacement of one end of the spring *relative* to the other. Similarly, the damping force in any damper

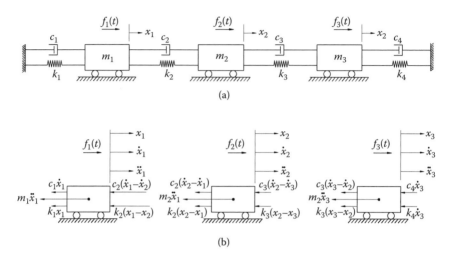

Figure 3.1 A three degree-of-freedom spring–mass system: (a) general arrangement, (b) free-body diagrams.

is equal to the damping coefficient of that damper multiplied by the velocity of one end of the damper (the piston) *relative* to the other end (the dashpot). The inertial force upon a given mass is, of course, simply given by the product of the mass and its *own* acceleration.

Writing down the equilibrium conditions for masses m_1, m_2 and m_3, we have, from the free-body diagrams,

$$m_1\ddot{x}_1 + k_1 x_1 + k_2(x_1 - x_2) + c_1\dot{x}_1 + c_2(\dot{x}_1 - \dot{x}_2) = f_1(t) \tag{3.1}$$

$$m_2\ddot{x}_2 + k_2(x_2 - x_1) + k_3(x_2 - x_3) + c_2(\dot{x}_2 - \dot{x}_1) + c_3(\dot{x}_2 - \dot{x}_3) = f_2(t) \tag{3.2}$$

$$m_3\ddot{x}_3 + k_3(x_3 - x_2) + k_4 x_3 + c_3(\dot{x}_3 - \dot{x}_2) + c_4\dot{x}_3 = f_3(t) \tag{3.3}$$

Rearranging the above equations, we obtain

$$m_1\ddot{x}_1 + (k_1 + k_2)x_1 - k_2 x_2 + (c_1 + c_2)\dot{x}_1 - c_2\dot{x}_2 = f_1(t) \tag{3.4}$$

$$m_2\ddot{x}_2 - k_2 x_1 + (k_2 + k_3)x_2 - k_3 x_3 - c_2\dot{x}_1 + (c_2 + c_3)\dot{x}_2 - c_3\dot{x}_3 = f_2(t) \tag{3.5}$$

$$m_3\ddot{x}_3 - k_3 x_2 + (k_3 + k_4)x_3 - c_3\dot{x}_2 + (c_3 + c_4)\dot{x}_3 = f_3(t) \tag{3.6}$$

In matrix form, the equations become

$$
\begin{bmatrix} m_1 & 0 & 0 \\ 0 & m_2 & 0 \\ 0 & 0 & m_3 \end{bmatrix} \begin{bmatrix} \ddot{x}_1 \\ \ddot{x}_2 \\ \ddot{x}_3 \end{bmatrix} + \begin{bmatrix} (c_1 + c_2) & -c_2 & 0 \\ -c_2 & (c_2 + c_3) & -c_3 \\ 0 & -c_3 & (c_3 + c_4) \end{bmatrix} \begin{bmatrix} \dot{x}_1 \\ \dot{x}_2 \\ \dot{x}_3 \end{bmatrix}
$$

$$
+ \begin{bmatrix} (k_1 + k_2) & -k_2 & 0 \\ -k_2 & (k_2 + k_3) & -k_3 \\ 0 & -k_3 & (k_3 + k_4) \end{bmatrix} \begin{bmatrix} x_1 \\ x_2 \\ x_3 \end{bmatrix} = \begin{bmatrix} f_1(t) \\ f_2(t) \\ f_3(t) \end{bmatrix} \tag{3.7}
$$

that is,

$$
M\ddot{X} + C\dot{X} + KX = F \tag{3.8}
$$

where M is called the *mass matrix*, C is called the *damping matrix*, and K is called the *stiffness matrix*; X, $\dot{X}$ and $\ddot{X}$ are the *displacement vector*, *velocity vector* and the *acceleration vector*, respectively; F is the *force vector*.

Note that the equations of motion are *coupled*, since each features more than one coordinate.

For discrete parameter models, the mass matrix M is a diagonal matrix, for which all elements are zero except those on the diagonal running from the top left to the bottom right of the matrix. The stiffness matrix K and the damping matrix C are both symmetric about the diagonal running from top left to bottom right.

For undamped free vibration, Equation (3.8) reduces to

$$
M\ddot{X} + KX = 0 \tag{3.9}
$$

3.3 TECHNIQUES FOR ASSEMBLING THE STIFFNESS MATRIX

Consider the undamped free vibration of a system with n degrees of freedom, for which Equation (3.9) may be written in the following expanded form:

$$
\begin{bmatrix} m_{11} & & & \\ & m_{22} & & \\ & & \cdot & \\ & & & m_{nn} \end{bmatrix} \begin{bmatrix} \ddot{x}_1 \\ \ddot{x}_2 \\ \cdot \\ \cdot \\ \ddot{x}_n \end{bmatrix} + \begin{bmatrix} k_{11} & k_{12} & . & . & k_{1n} \\ k_{21} & k_{22} & . & . & k_{2n} \\ . & . & . & . & . \\ . & . & . & . & . \\ k_{n1} & k_{n2} & . & . & k_{nn} \end{bmatrix} \begin{bmatrix} x_1 \\ x_2 \\ \cdot \\ \cdot \\ x_n \end{bmatrix} = \begin{bmatrix} 0 \\ 0 \\ \cdot \\ \cdot \\ 0 \end{bmatrix}
$$

$$
\tag{3.10}
$$

Let us assume that the system is given a unit displacement at the jth coordinate (that is, $x_j = 1$), while all the remaining $n - 1$ coordinates are held at zero. In order to maintain this configuration of the system, forces will be required at the various locations of the coordinates. In the matrix K, the element k_{ij}, called a *stiffness coefficient*, is the force required at coordinate x_i when a unit displacement is assigned to coordinate x_j, while all coordinates other than x_j are held at zero.

As an example, consider the four-storey shear frame shown in Figure 3.2, which has four degrees of freedom $\{x_1, x_2, x_3, x_4\}$, these being the lateral sways of the slab masses $\{m_1, m_2, m_3, m_4\}$. The stiffnesses of the

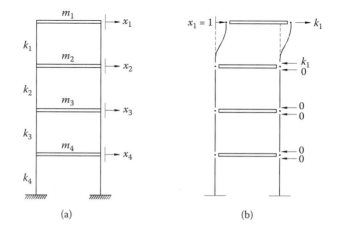

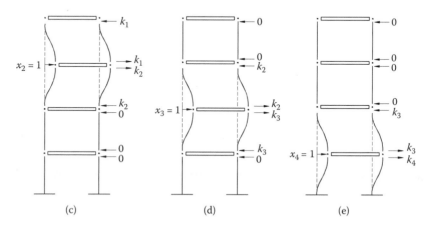

Figure 3.2 Determination of stiffness coefficients illustrated by a four-storey shear frame: (a) general arrangement, (b) application of $x_1 = 1$, (c) application of $x_2 = 1$, (d) application of $x_3 = 1$, (e) application of $x_4 = 1$. In (b–e), lateral forces on column ends, required to maintain the indicated deformation configurations, are shown.

column intervals are, from the top downward, $\{k_1, k_2, k_3, k_4\}$, as shown in Figure 3.2(a). The bases of the columns are assumed to be fully restrained against rotation, as are the column junctions to the horizontal slabs.

Figure 3.2(b–e) shows, in free-body format, the lateral forces on each slab mass, exerted by the columns with which it is in contact, when the shear frame is assigned the displacement sets $\{x_1 = 1, x_2 = 0, x_3 = 0, x_4 = 0\}$, $\{x_1 = 0, x_2 = 1, x_3 = 0, x_4 = 0\}$, $\{x_1 = 0, x_2 = 0, x_3 = 1, x_4 = 0\}$, $\{x_1 = 0, x_2 = 0, x_3 = 0, x_4 = 1\}$, respectively.

As a result of unit displacement of mass m_1 while all the other masses are held at their equilibrium positions (Figure 3.2(b)), the net forces acting upon masses m_1, m_2, m_3 and m_4 are k_1, $-k_1$, 0 and 0, respectively. That is, $k_{11} = k_1$, $k_{21} = -k_1$, $k_{31} = 0$ and $k_{41} = 0$.

Similarly, as a result of unit displacement of mass m_2 while all the other masses are held at their equilibrium positions (Figure 3.2(c)), the net forces acting upon masses m_1, m_2, m_3 and m_4 are $-k_1$, $(k_1 + k_2)$, $-k_2$ and 0, respectively. That is, $k_{12} = -k_1$, $k_{22} = (k_1 + k_2)$, $k_{32} = -k_2$ and $k_{42} = 0$.

From Figure 3.2(d, e), the rest of the stiffness coefficients are as follows:

$$k_{13} = 0; \; k_{23} = -k_2; \; k_{33} = (k_2 + k_3); \; k_{43} = -k_3$$

$$k_{14} = 0; \; k_{24} = 0; \; k_{34} = -k_3; \; k_{44} = (k_3 + k_4)$$

Putting all the above results together, we obtain

$$K = \begin{bmatrix} k_1 & -k_1 & 0 & 0 \\ -k_1 & (k_1 + k_2) & -k_2 & 0 \\ 0 & -k_2 & (k_2 + k_3) & -k_3 \\ 0 & 0 & -k_3 & (k_3 + k_4) \end{bmatrix} \tag{3.11}$$

The mass matrix of the system is simply

$$M = \begin{bmatrix} m_1 & 0 & 0 & 0 \\ 0 & m_2 & 0 & 0 \\ 0 & 0 & m_3 & 0 \\ 0 & 0 & 0 & m_4 \end{bmatrix} \tag{3.12}$$

As a second example, we consider a three degree-of-freedom spring–mass system, with a general arrangement as shown in Figure 3.3(a). To generate the stiffness coefficients, we apply unit values of displacements x_j ($j = 1$ in Figure 3.3(b), $j = 2$ in Figure 3.3(c), $j = 3$ in Figure 3.3(d)), while all the other displacements are held at zero, and note the forces required at each mass

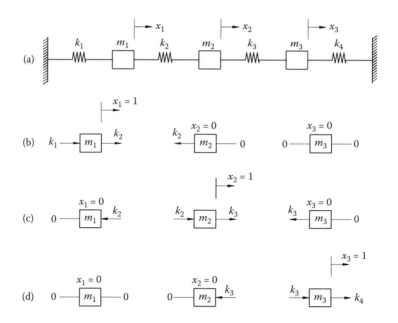

Figure 3.3 Determination of stiffness coefficients for a three degree-of-freedom spring–mass system: (a) general arrangement, (b–d) free-body diagrams for the application of $x_1 = 1$, $x_2 = 1$ and $x_3 = 1$, respectively.

in order to maintain each of the three configurations. From the free-body diagrams in Figure 3.3(b–d), we read off the nine stiffness coefficients as the net forces acting upon each of the three masses m_1, m_2 and m_3, for each of the three deformation configurations:

$$k_{11} = (k_1 + k_2); \ k_{21} = -k_2; \ k_{31} = 0$$

$$k_{12} = -k_2; \ k_{22} = (k_2 + k_3); \ k_{32} = -k_3$$

$$k_{13} = 0; \ k_{23} = -k_3; \ k_{33} = (k_3 + k_4)$$

The stiffness matrix K is therefore

$$K = \begin{bmatrix} (k_1 + k_2) & -k_2 & 0 \\ -k_2 & (k_2 + k_3) & -k_3 \\ 0 & -k_3 & (k_3 + k_4) \end{bmatrix} \quad (3.13)$$

which is the same result as was obtained in the last section by writing down the equations of motion from first principles.

The mass matrix of the system is simply

$$M = \begin{bmatrix} m_1 & 0 & 0 \\ 0 & m_2 & 0 \\ 0 & 0 & m_3 \end{bmatrix} \qquad (3.14)$$

In the case of the shaft–disc torsional system shown in Figure 3.4, the degrees of freedom are now the rotations θ_1, θ_2 and θ_3 of the discs I_1, I_2 and I_3 (the I's refer to the moments of inertia of the discs about their common axis of rotation). The stiffnesses k_1, k_2, k_3 and k_4 are now torsional stiffnesses of the shaft intervals, given by $k = GJ/l$, where G is the shear modulus of the shaft material, J the polar second moment of area of the cross section of a shaft interval, and l the length of the shaft interval.

Applying unit rotations θ_j ($j = 1,2,3$ in turn), with all rotations other than θ_j suppressed, we obtain the stiffness coefficients as the net torque upon each of the three discs, required to maintain each of the three configurations. Clearly, K is of exactly the same form as given by Equation (3.13), while the mass matrix M now consists of the polar moments of inertia:

$$M = \begin{bmatrix} I_1 & 0 & 0 \\ 0 & I_2 & 0 \\ 0 & 0 & I_3 \end{bmatrix} \qquad (3.15)$$

In full, the equation of motion for the undamped free vibration of this system reads

$$\begin{bmatrix} I_1 & 0 & 0 \\ 0 & I_2 & 0 \\ 0 & 0 & I_3 \end{bmatrix} \begin{bmatrix} \ddot{\theta}_1 \\ \ddot{\theta}_2 \\ \ddot{\theta}_3 \end{bmatrix} + \begin{bmatrix} (k_1 + k_2) & -k_2 & 0 \\ -k_2 & (k_2 + k_3) & -k_3 \\ 0 & -k_3 & (k_3 + k_4) \end{bmatrix} \begin{bmatrix} \theta_1 \\ \theta_2 \\ \theta_3 \end{bmatrix} = \begin{bmatrix} 0 \\ 0 \\ 0 \end{bmatrix}$$

$$(3.16)$$

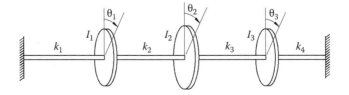

Figure 3.4 A three degree-of-freedom shaft–disc system.

that is,

$$M\ddot{X} + KX = 0 \tag{3.17}$$

where, in this example, X and $\ddot{X}$ are rotation vectors and angular acceleration vectors, respectively.

3.4 THE FLEXIBILITY FORMULATION OF THE EQUATIONS OF MOTION AND ASSEMBLY OF THE FLEXIBILITY MATRIX

Sometimes it is more convenient to obtain the *flexibility coefficients* of a
system rather than its stiffness coefficients. If we premultiply Equation (3.9)
by the inverse of the stiffness matrix, we obtain

$$K^{-1}M\ddot{X} + K^{-1}KX = 0 \tag{3.18}$$

that is,

$$DM\ddot{X} + X = 0 \tag{3.19}$$

since $K^{-1} = D$, the flexibility matrix, and $K^{-1}K$ is, of course, an identity
matrix. Writing out Equation (3.19) in expanded form, we have, for a system with n degrees of freedom,

$$\begin{bmatrix} d_{11} & d_{12} & . & . & d_{1n} \\ d_{21} & d_{22} & . & . & d_{2n} \\ . & . & . & . & . \\ . & . & . & . & . \\ d_{n1} & d_{n2} & . & . & d_{nn} \end{bmatrix} \begin{bmatrix} m_{11} & & & \\ & m_{22} & & \\ & & . & \\ & & & m_{nn} \end{bmatrix} \begin{bmatrix} \ddot{x}_1 \\ \ddot{x}_2 \\ . \\ . \\ \ddot{x}_n \end{bmatrix} + \begin{bmatrix} x_1 \\ x_2 \\ . \\ . \\ x_n \end{bmatrix} = \begin{bmatrix} 0 \\ 0 \\ . \\ . \\ 0 \end{bmatrix}$$

$$\tag{3.20}$$

Let us assume that a unit force is applied at the location of coordinate x_j,
while the locations of all the remaining $n - 1$ coordinates are not subjected
to any force. In the flexibility matrix D, the element d_{ij} is the displacement
that ensues at coordinate x_i as a result of a unit force applied at the location
of coordinate x_j, while all the locations of coordinates other than x_j are
unloaded.

Consider a simply supported beam of negligible self-weight and constant flexural rigidity EI, carrying three concentrated masses m_1, m_2

and m_3, with degrees of freedom in the form of small vertical motions x_1, x_2 and x_3, respectively, as shown in Figure 3.5(a). For this system, the mass matrix is simply

$$M = \begin{bmatrix} m_1 & 0 & 0 \\ 0 & m_2 & 0 \\ 0 & 0 & m_3 \end{bmatrix} \tag{3.21}$$

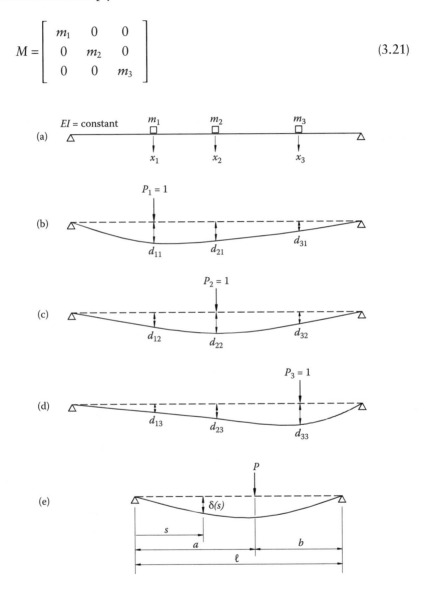

Figure 3.5 Determination of flexibility coefficients for a simply supported beam carrying three concentrated masses: (a) general arrangement, (b–d) deflection diagrams for the application of $P_1 = 1$, $P_2 = 1$ and $P_3 = 1$, respectively, (e) deflection at an arbitrary location due to an arbitrarily positioned vertical point load P.

To obtain the flexibility matrix, we first apply a unit vertical force at the location of x_1, and note the vertical deflections that ensue at the locations of x_1, x_2 and x_3; these are numerically equal to d_{11}, d_{21} and d_{31}, respectively, as shown in Figure 3.5(b). Similarly, application of a unit vertical force at the location of x_2 yields $\{d_{12}, d_{22}, d_{32}\}$, as shown in Figure 3.5(c), while application of a unit vertical force at the location of x_3 yields $\{d_{13}, d_{23}, d_{33}\}$, as shown in Figure 3.5(d). These static deflections may easily be calculated using standard formulae of structural analysis. With reference to Figure 3.5(e), the vertical deflection $\delta(s)$ at a distance s from the left-hand end of a simply supported beam of constant flexural rigidity EI and length l, due to a vertical point load P positioned at a distance of a from the left-hand end of the beam, is given by

$$\delta(s) = \frac{Pbs}{6EIl}\left[a(b+l) - s^2\right] \text{ for } s \leq a \tag{3.22}$$

$$\delta(s) = \frac{Pa(l-s)}{6EIl}\left[s(2l-s) - a^2\right] \text{ for } s \geq a \tag{3.23}$$

where $b = l - a$.

Another instance when it may be simpler to adopt the flexibility formulation is the case of small transverse vibrations of a taut light cable carrying concentrated masses. The cable in Figure 3.6(a) is highly tensioned by a prestress force T, which is constant over the entire length of the cable, and does not significantly change during small vertical motions of four concentrated masses m_1, m_2, m_3 and m_4 carried by the cable, itself assumed to be weightless. Figure 3.6(b–e) shows the displacements at the locations of all four degrees of freedom $\{x_1, x_2, x_3, x_4\}$, when a unit vertical force is applied at the locations of x_1, x_2, x_3 and x_4, in turn. By definition, these displacements are the flexibility coefficients d_{ij}, as shown. Note that the taut cable remains straight between either anchorage point and the loaded point.

The computation of the d_{ij}'s is illustrated by reference to Figure 3.7, which shows a taut cable carrying a prestress tension T, and subjected to a vertical point load $P = 1$ at a distance a from the left end and b from the right end. Assume this unit vertical load is applied at the location of coordinate x_j in the dynamics problem of Figure 3.6. To find the vertical deflection d_{jj} under the point load, we consider vertical equilibrium there, which, from Figure 3.7(a), requires that

$$T\sin\alpha + T\sin\beta = 1 \tag{3.24}$$

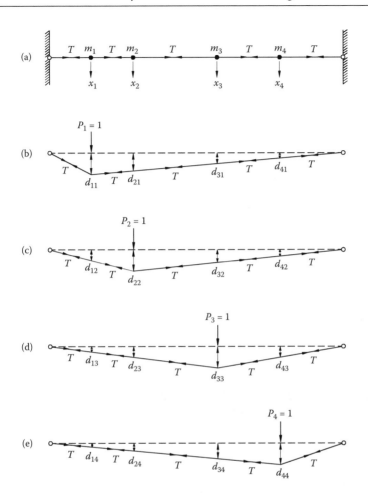

Figure 3.6 Determination of flexibility coefficients for a taut weightless cable carrying four concentrated masses: (a) general arrangement, (b–e) deflection diagrams for the application of $P_1 = 1$, $P_2 = 1$, $P_3 = 1$ and $P_4 = 1$, respectively.

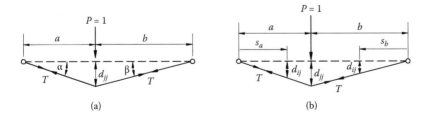

Figure 3.7 Computation of the d_{ij} of Figure 3.6: (a) displacement under the point load, (b) displacements away from the point load.

Since angles α and β are very small, the above equation may be expressed as

$$T\left(\frac{d_{jj}}{a} + \frac{d_{jj}}{b}\right) = 1 \tag{3.25}$$

from which we obtain

$$d_{jj} = \frac{1}{T}\left(\frac{ab}{a+b}\right) \tag{3.26}$$

Once the displacement d_{jj} under the point load (that is, at the location of x_j) is known, the displacements at the locations of coordinates x_i $(i \neq j)$ follow from considerations of similar triangles. By reference to Figure 3.7(b), we have

$$d_{ij} = \frac{s_a}{a}\,d_{jj} \tag{3.27}$$

for locations to the left of P, and

$$d_{ij} = \frac{s_b}{b}\,d_{jj} \tag{3.28}$$

for locations to the right of P.

3.5 DETERMINATION OF NATURAL FREQUENCIES AND MODE SHAPES

From relation (2.13), which for the case of underdamping reads $\omega_d = \omega_n\sqrt{1-\xi^2}$, we observe that the damped natural circular frequency ω_d of a single degree-of-freedom system is nearly the same as the undamped natural circular frequency ω_n in many real situations with a relatively low damping factor ξ (say, $\xi \leq 0.25$). We may assume the same to hold for multi-degree-of-freedom systems, and proceed on the basis of the undamped free vibration equations of motion, thereby considerably simplifying the derivations without too much loss in accuracy.

As for a single degree-of-freedom system, we assume harmonic motion for every mass of a system with n degrees of freedom:

$$\begin{bmatrix} x_1 \\ x_2 \\ . \\ . \\ x_n \end{bmatrix} = \begin{bmatrix} X_1\cos(\omega t - \alpha) \\ X_2\cos(\omega t - \alpha) \\ . \\ . \\ X_n\cos(\omega t - \alpha) \end{bmatrix} \tag{3.29}$$

that is,

$$X = \phi \cos(\omega t - \alpha) \tag{3.30}$$

where

$$\phi = \begin{bmatrix} X_1 \\ X_2 \\ . \\ . \\ X_n \end{bmatrix} \tag{3.31}$$

From Equation (3.30), we have

$$\ddot{X} = -\phi\, \omega^2 \cos(\omega t - \alpha) \tag{3.32}$$

Substituting for $\ddot{X}$ and X in the equation of motion (that is, Equation (3.9)), we obtain

$$-M\phi\, \omega^2 \cos(\omega t - \alpha) + K\phi \cos(\omega t - \alpha) = 0 \tag{3.33}$$

Dividing throughout by $\cos(\omega t - \alpha)$ and rearranging, we obtain

$$(K - \omega^2 M)\,\phi = 0 \tag{3.34}$$

which is an *eigenvalue equation*. Let us introduce the notation $\lambda = \omega^2$ to denote an *eigenvalue* of the problem. The eigenvalue equation becomes

$$(K - \lambda M)\,\phi = 0 \tag{3.35}$$

For nontrivial solutions of Equation (3.35) to exist, then the determinant of $(K - \lambda M)$ must be zero, that is,

$$|K - \lambda M| = 0 \tag{3.36}$$

In expanded form, this condition becomes

$$\begin{vmatrix} (k_{11} - \lambda m_{11}) & k_{12} & . & . & k_{1n} \\ k_{21} & (k_{22} - \lambda m_{22}) & . & . & k_{2n} \\ . & . & . & . & . \\ . & . & . & . & . \\ k_{n1} & k_{n2} & . & . & (k_{nn} - \lambda m_{nn}) \end{vmatrix} = 0 \tag{3.37}$$

The evaluation of this determinant leads to a polynomial equation of nth degree in λ, called the *characteristic equation* of the system. The n roots of the characteristic equation, namely, $\lambda_1(=\omega_1^2)$, $\lambda_2(=\omega_2^2)$, ..., $\lambda_n(=\omega_n^2)$, are the eigenvalues of the system, from which the natural circular frequencies follow.

Substituting a particular eigenvalue λ_r into the eigenvalue equation (that is, Equation (3.35)), we obtain

$$
\begin{bmatrix}
(k_{11} - \lambda_r m_{11}) & k_{12} & . & . & k_{1n} \\
k_{21} & (k_{22} - \lambda_r m_{22}) & . & . & k_{2n} \\
. & . & . & . & . \\
. & . & . & . & . \\
k_{n1} & k_{n2} & . & . & (k_{nn} - \lambda_r m_{nn})
\end{bmatrix}
\begin{bmatrix}
X_1 \\
X_2 \\
. \\
. \\
X_n
\end{bmatrix}_r
=
\begin{bmatrix}
0 \\
0 \\
. \\
. \\
0
\end{bmatrix}
\tag{3.38}
$$

where the vector

$$
\begin{bmatrix}
X_1 \\
X_2 \\
. \\
. \\
X_n
\end{bmatrix}_r
= \phi_r
\tag{3.39}
$$

is the eigenvector corresponding to the natural circular frequency ω_r. It is only possible to solve the above homogeneous algebraic equation in relative form (for instance, we can express X_2, X_3, ..., X_n in terms of X_1).

Thus, an eigenvector ϕ_r expresses the relative values of the amplitudes $\{X_1, X_2, ..., X_n\}$ for vibration at the natural frequency ω_r. It describes the shape of a particular mode of vibration (corresponding to the frequency ω_r), called a *mode shape*. There are as many mode shapes as there are frequencies.

3.6 THE FLEXIBILITY FORMULATION OF THE EIGENVALUE PROBLEM

As before, we assume harmonic motion for every mass of the system, and substitute expressions (3.30) and (3.32) for X and $\ddot{X}$, respectively, into the flexibility equation of motion (that is, Equation (3.19)). This yields

$$
\left[\left(-DM\omega^2 + I\right)\phi\right]\cos(\omega t - \alpha) = 0
\tag{3.40}
$$

which, after replacing ω^2 with λ, leads to the eigenvalue equation

$$(I - \lambda DM)\phi = 0 \tag{3.41}$$

and the characteristic equation

$$|I - \lambda DM| = 0 \tag{3.42}$$

where I is, of course, the $n \times n$ identity matrix.

3.7 WORKED EXAMPLES

Example 3.1
Figure 3.8(a) shows a three-storey shear frame with degrees of freedom x_1, x_2 and x_3 corresponding to slab masses m, $2m$ and $2m$, respectively. Moving from the top towards the bottom, the column interval stiffnesses are $2k$, $3k$ and $4k$, as shown. It is required to obtain the natural frequencies and mode shapes of the system for undamped free vibration.
The mass matrix for the system is

$$M = \begin{bmatrix} m & 0 & 0 \\ 0 & 2m & 0 \\ 0 & 0 & 2m \end{bmatrix}$$

From Figure 3.8(b–d), we obtain the stiffness matrix as follows:

$$K = \begin{bmatrix} 2k & -2k & 0 \\ -2k & 5k & -3k \\ 0 & -3k & 7k \end{bmatrix}$$

For undamped free vibration, $|K - \lambda M| = 0$, that is,

$$\begin{vmatrix} (2k - \lambda m) & -2k & 0 \\ -2k & (5k - 2\lambda m) & -3k \\ 0 & -3k & (7k - 2\lambda m) \end{vmatrix} = 0$$

This results in the characteristic equation

$$4m^3\lambda^3 - 32m^2 k\lambda^2 + 66mk^2\lambda - 24k^3 = 0$$

which, upon solving, yields the roots

$$\lambda_1 = 0.461\frac{k}{m}; \ \lambda_2 = 2.682\frac{k}{m}; \ \lambda_3 = 4.857\frac{k}{m}$$

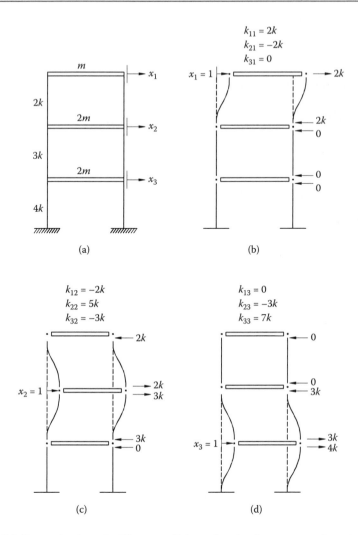

Figure 3.8 Determination of stiffness coefficients for the three-storey shear frame of Example 3.1: (a) general arrangement, (b–d) free-body diagrams for the application of $x_1 = 1$, $x_2 = 1$ and $x_3 = 1$, respectively.

from which the natural circular frequencies are obtained as

$$\omega_1 = 0.679\sqrt{\frac{k}{m}}; \quad \omega_2 = 1.638\sqrt{\frac{k}{m}}; \quad \omega_3 = 2.204\sqrt{\frac{k}{m}}$$

and the natural frequencies as

$$h_1 = \frac{\omega_1}{2\pi} = 0.108\sqrt{\frac{k}{m}}; \quad h_2 = \frac{\omega_2}{2\pi} = 0.261\sqrt{\frac{k}{m}}; \quad h_3 = \frac{\omega_3}{2\pi} = 0.351\sqrt{\frac{k}{m}}$$

To determine the mode shapes corresponding to these three natural frequencies, we substitute λ_1, λ_2 and λ_3 in turn into the eigenvalue equation $(K - \lambda M)\phi = 0$. Taking $\lambda = \lambda_1 = 0.461k/m$, we have

$$
\begin{bmatrix}
2k - \left(0.461\dfrac{k}{m}\right)m & -2k & 0 \\[2ex]
-2k & 5k - 2\left(0.461\dfrac{k}{m}\right)m & -3k \\[2ex]
0 & -3k & 7k - 2\left(0.461\dfrac{k}{m}\right)m
\end{bmatrix}
\begin{bmatrix} X_1 \\ X_2 \\ X_3 \end{bmatrix}_1
=
\begin{bmatrix} 0 \\ 0 \\ 0 \end{bmatrix}
$$

that is,

$$
\begin{bmatrix}
1.539 & -2 & 0 \\
-2 & 4.078 & -3 \\
0 & -3 & 6.078
\end{bmatrix}
\begin{bmatrix} X_1 \\ X_2 \\ X_3 \end{bmatrix}_1
=
\begin{bmatrix} 0 \\ 0 \\ 0 \end{bmatrix}
$$

From the first equation of the above matrix set, we have

$$1.539X_1 - 2X_2 = 0$$

from which

$$X_2 = 0.770X_1$$

From the second equation of the matrix set, we have

$$-2X_1 + 4.078X_2 - 3X_3 = 0$$

Using the above value for X_2, this expression becomes

$$-2X_1 + 4.078(0.770X_1) - 3X_3 = 0$$

from which

$$X_3 = 0.380X_1$$

From the third equation of the matrix set, we have

$$-3X_2 + 6.078X_3 = 0$$

Using the obtained values for X_2 and X_3, this expression becomes

$$-3(0.770X_1) + 6.078(0.380X_1) = 0$$

that is,

$$-2.310X_1 + 2.310X_1 = 0$$

which, as expected, is identically satisfied.

The eigenvector corresponding to ω_1 is therefore

$$\phi_1 = \begin{bmatrix} 1.000 \\ 0.770 \\ 0.380 \end{bmatrix} X_1$$

Next, taking $\lambda = \lambda_2 = 2.682k/m$, we have

$$\begin{bmatrix} 2k-\left(2.682\dfrac{k}{m}\right)m & -2k & 0 \\ -2k & 5k-2\left(2.682\dfrac{k}{m}\right)m & -3k \\ 0 & -3k & 7k-2\left(2.682\dfrac{k}{m}\right)m \end{bmatrix} \begin{bmatrix} X_1 \\ X_2 \\ X_3 \end{bmatrix}_2 = \begin{bmatrix} 0 \\ 0 \\ 0 \end{bmatrix}$$

that is,

$$\begin{bmatrix} -0.682 & -2 & 0 \\ -2 & -0.364 & -3 \\ 0 & -3 & 1.636 \end{bmatrix} \begin{bmatrix} X_1 \\ X_2 \\ X_3 \end{bmatrix}_2 = \begin{bmatrix} 0 \\ 0 \\ 0 \end{bmatrix}$$

From the first equation of the above matrix set, we have

$$-0.682X_1 - 2X_2 = 0$$

from which

$$X_2 = -0.341X_1$$

From the second equation of the matrix set, we have

$$-2X_1 - 0.364X_2 - 3X_3 = 0$$

Using the obtained value for X_2, this expression becomes

$$-2X_1 - 0.364(-0.341X_1) - 3X_3 = 0$$

from which

$$X_3 = -0.625X_1$$

From the third equation of the matrix set, we have

$$-3X_2 + 1.636X_3 = 0$$

Using the obtained values for X_2 and X_3, this expression becomes

$$-3(-0.341X_1) + 1.636(-0.625X_1) = 0$$

that is,

$$1.023X_1 - 1.023X_1 = 0$$

which, as expected, is identically satisfied.
   The eigenvector corresponding to ω_2 is therefore

$$\phi_2 = \begin{bmatrix} 1.000 \\ -0.341 \\ -0.625 \end{bmatrix} X_1$$

Finally, taking $\lambda = \lambda_3 = 4.857\,k/m$, we have

$$\begin{bmatrix} 2k - \left(4.857\dfrac{k}{m}\right)m & -2k & 0 \\ -2k & 5k - 2\left(4.857\dfrac{k}{m}\right)m & -3k \\ 0 & -3k & 7k - 2\left(4.857\dfrac{k}{m}\right)m \end{bmatrix} \begin{bmatrix} X_1 \\ X_2 \\ X_3 \end{bmatrix}_3 = \begin{bmatrix} 0 \\ 0 \\ 0 \end{bmatrix}$$

that is,

$$\begin{bmatrix} -2.857 & -2 & 0 \\ -2 & -4.714 & -3 \\ 0 & -3 & -2.714 \end{bmatrix} \begin{bmatrix} X_1 \\ X_2 \\ X_3 \end{bmatrix}_3 = \begin{bmatrix} 0 \\ 0 \\ 0 \end{bmatrix}$$

From the first equation,

$$-2.857X_1 - 2X_2 = 0$$

we obtain

$$X_2 = -1.429X_1$$

The second equation, with X_2 as obtained above, becomes

$$-2X_1 - 4.714(-1.429X_1) - 3X_3 = 0$$

from which

$$X_3 = 1.579X_1$$

The third equation, with X_2 and X_3 as obtained above, reads

$$-3(-1.429X_1) - 2.714(1.579X_1) = 0$$

that is,

$$4.287X_1 - 4.285X_1 = 0$$

which, disregarding rounding-off errors, is identically satisfied, as expected.

The eigenvector corresponding to ω_3 is therefore

$$\phi_3 = \begin{bmatrix} 1.000 \\ -1.429 \\ 1.579 \end{bmatrix} X_1$$

The mode shapes corresponding to ω_1, ω_2 and ω_3 are plotted in Figure 3.9.

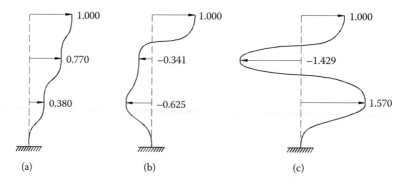

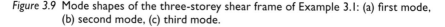

Figure 3.9 Mode shapes of the three-storey shear frame of Example 3.1: (a) first mode, (b) second mode, (c) third mode.

Example 3.2
Figure 3.10(a) shows a simply supported beam of length l and constant flexural rigidity EI. It carries concentrated masses m and $3m$ at distances $l/4$ and $3l/4$, respectively, from the left end. The masses m and $3m$ have degrees of freedom x_1 and x_2, which are vertical motions as the beam undergoes small transverse vibrations. Assuming the beam itself is of negligible mass, determine the natural frequencies and mode shapes of the system for undamped free vibration.

The mass matrix for the system is simply

$$M = \begin{bmatrix} m & 0 \\ 0 & 3m \end{bmatrix}$$

In this example, it is easier to proceed on the basis of the flexibility formulation. By reference to Figure 3.10(b, c), and the formulae

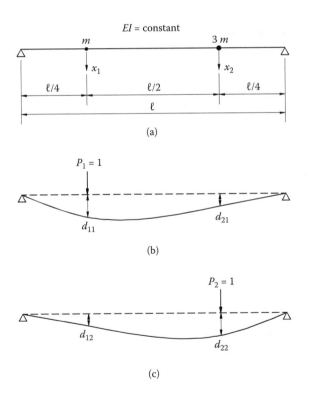

Figure 3.10 Determination of flexibility coefficients for the simply supported beam of Example 3.2: (a) general arrangement, (b) deflection diagram for the application of $P_1 = 1$, (c) deflection diagram for the application of $P_2 = 1$.

of Equations (3.22) and (3.23), the four flexibility coefficients are obtained as follows:

$$d_{11} = \frac{(1)\left(\frac{3l}{4}\right)\left(\frac{l}{4}\right)}{6EIl}\left[\left(\frac{l}{4}\right)\left(\frac{7l}{4}\right) - \left(\frac{l}{4}\right)^2\right] = \frac{3l^3}{256EI}$$

$$d_{21} = \frac{(1)\left(\frac{l}{4}\right)\left(\frac{l}{4}\right)}{6EIl}\left[\left(\frac{3l}{4}\right)\left(\frac{5l}{4}\right) - \left(\frac{l}{4}\right)^2\right] = \frac{7l^3}{256 \times 3EI}$$

$$d_{12} = d_{21}$$

$$d_{22} = d_{11}$$

The equality of d_{12} and d_{21} stems from the general property $d_{ij} = d_{ji}$ of the flexibility matrix, which in turn stems from the reciprocal theorem. On the other hand, the equality of d_{11} and d_{22} stems from the fact that the distance of x_1 from the left support is the same as the distance of x_2 from the right support.

Thus,

$$D = \frac{l^3}{768EI}\begin{bmatrix} 9 & 7 \\ 7 & 9 \end{bmatrix} = \begin{bmatrix} 9r & 7r \\ 7r & 9r \end{bmatrix}$$

where

$$r = \frac{l^3}{768EI}$$

For undamped free vibration, $|I - \lambda DM| = 0$, that is,

$$\left\| \begin{bmatrix} 1 & 0 \\ 0 & 1 \end{bmatrix} - \lambda \begin{bmatrix} 9r & 7r \\ 7r & 9r \end{bmatrix} \begin{bmatrix} m & 0 \\ 0 & 3m \end{bmatrix} \right\| = 0$$

This simplifies to

$$\begin{vmatrix} (1 - 9rm\lambda) & -21rm\lambda \\ -7rm\lambda & (1 - 27rm\lambda) \end{vmatrix} = 0$$

leading to the characteristic equation

$$96m^2r^2\lambda^2 - 36mr\lambda + 1 = 0$$

with roots

$$\lambda_1 = 0.03021\left(\frac{1}{rm}\right); \quad \lambda_2 = 0.34479\left(\frac{1}{rm}\right)$$

Therefore, the natural circular frequencies are

$$\omega_1 = 0.17381\sqrt{\frac{1}{rm}} \; ; \; \omega_2 = 0.58719\sqrt{\frac{1}{rm}}$$

while the actual frequencies (in cycles per second) are

$$h_1 = \frac{\omega_1}{2\pi} = 0.02766\sqrt{\frac{1}{rm}}; \; h_2 = \frac{\omega_2}{2\pi} = 0.09345\sqrt{\frac{1}{rm}}$$

Substituting $\lambda = \lambda_1 = 0.03021(1/rm)$ into the eigenvalue equation $(I - \lambda DM)\phi = 0$, we obtain

$$\begin{bmatrix} 1-9rm\left(\dfrac{0.03021}{rm}\right) & -21rm\left(\dfrac{0.03021}{rm}\right) \\ -7rm\left(\dfrac{0.03021}{rm}\right) & 1-27rm\left(\dfrac{0.03021}{rm}\right) \end{bmatrix} \begin{bmatrix} X_1 \\ X_2 \end{bmatrix}_1 = \begin{bmatrix} 0 \\ 0 \end{bmatrix}$$

that is,

$$\begin{bmatrix} 0.7281 & -0.6344 \\ -0.2115 & 0.1843 \end{bmatrix} \begin{bmatrix} X_1 \\ X_2 \end{bmatrix}_1 = \begin{bmatrix} 0 \\ 0 \end{bmatrix}$$

From the first equation, we have

$$0.7281X_1 - 0.6344X_2 = 0$$

from which

$$X_2 = 1.148X_1$$

From the second equation, we have

$$-0.2115X_1 + 0.1843X_2 = 0$$

from which

$$X_2 = 1.148X_1$$

which, as expected, is the same result as yielded by the first equation. The eigenvector corresponding to ω_1 is therefore

$$\phi_1 = \begin{bmatrix} 1.000 \\ 1.148 \end{bmatrix} X_1$$

Taking $\lambda = \lambda_2 = 0.34479(1/rm)$, we have

$$\begin{bmatrix} 1-9rm\left(\dfrac{0.34479}{rm}\right) & -21rm\left(\dfrac{0.34479}{rm}\right) \\ -7rm\left(\dfrac{0.34479}{rm}\right) & 1-27rm\left(\dfrac{0.34479}{rm}\right) \end{bmatrix} \begin{bmatrix} X_1 \\ X_2 \end{bmatrix}_2 = \begin{bmatrix} 0 \\ 0 \end{bmatrix}$$

that is,

$$\begin{bmatrix} -2.1031 & -7.2406 \\ -2.4135 & -8.3093 \end{bmatrix} \begin{bmatrix} X_1 \\ X_2 \end{bmatrix}_2 = \begin{bmatrix} 0 \\ 0 \end{bmatrix}$$

From the first equation, we have

$$-2.1031X_1 - 7.2406X_2 = 0 \Rightarrow X_2 = -0.290X_1$$

From the second equation, we have

$$-2.4135X_1 - 8.3093X_2 = 0 \Rightarrow X_2 = -0.290X_1$$

which, of course, is the same result as yielded by the first equation. The eigenvector corresponding to ω_2 is therefore

$$\phi_2 = \begin{bmatrix} 1.000 \\ -0.290 \end{bmatrix} X_1$$

The mode shapes corresponding to ω_1 and ω_2 are plotted in Figure 3.11.

3.8 THE MODAL MATRIX

For a system with n degrees of freedom, the *modal matrix* Φ is the $n \times n$ square matrix obtained by collecting together all the n eigenvectors, as follows:

$$\Phi = [\phi_1 \ \phi_2 \ ... \ \phi_n]$$

$$= \begin{bmatrix} \begin{bmatrix} X_1 \\ X_2 \\ \cdot \\ \cdot \\ X_n \end{bmatrix}_1 & \begin{bmatrix} X_1 \\ X_2 \\ \cdot \\ \cdot \\ X_n \end{bmatrix}_2 & \cdot \ \cdot & \begin{bmatrix} X_1 \\ X_2 \\ \cdot \\ \cdot \\ X_n \end{bmatrix}_n \end{bmatrix} \qquad (3.43)$$

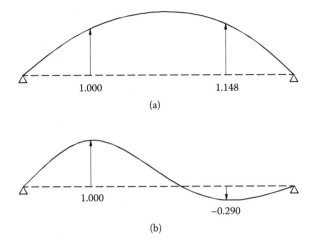

Figure 3.11 Mode shapes of the beam–mass system of Example 3.2: (a) first mode, (b) second mode.

For the first example of Section 3.7, we have the modal matrix

$$\Phi = \begin{bmatrix} 1.000 & 1.000 & 1.000 \\ 0.770 & -0.341 & -1.429 \\ 0.380 & -0.625 & 1.579 \end{bmatrix}$$

while for the second example of that section, the modal matrix reads

$$\Phi = \begin{bmatrix} 1.000 & 1.000 \\ 1.148 & -0.290 \end{bmatrix}$$

3.9 ORTHOGONALITY OF EIGENVECTORS

From Equation (3.34), we may write

$$K\phi = \omega^2 M\phi \qquad (3.44)$$

This relation holds for all modes of undamped free vibration of a system with two or more degrees of freedom. For the rth mode, we have

$$K\phi_r = \omega_r^2 M\phi_r \qquad (3.45)$$

while for the sth mode, we have

$$K\phi_s = \omega_s^2 M\phi_s \tag{3.46}$$

Premultiplying Equation (3.45) by ϕ_s^T, we obtain

$$\phi_s^T K\phi_r = \omega_r^2 \phi_s^T M\phi_r \tag{3.47}$$

Premultiplying Equation (3.46) by ϕ_r^T, we obtain

$$\phi_r^T K\phi_s = \omega_s^2 \phi_r^T M\phi_s \tag{3.48}$$

Transposing Equation (3.48) gives us

$$\left[\left(\phi_r^T K\right)\phi_s\right]^T = \omega_s^2 \left[\left(\phi_r^T M\right)\phi_s\right]^T \tag{3.49}$$

Using the well-known result $[AB]^T = B^T A^T$ of matrix algebra, we find that Equation (3.49) transforms to

$$\phi_s^T \left(\phi_r^T K\right)^T = \omega_s^2 \phi_s^T \left(\phi_r^T M\right)^T \tag{3.50}$$

which in turn becomes

$$\phi_s^T K^T \phi_r = \omega_s^2 \phi_s^T M^T \phi_r \tag{3.51}$$

Since K and M are symmetric square matrices, we have $K^T = K$ and $M^T = M$, so that the above equation becomes

$$\phi_s^T K\phi_r = \omega_s^2 \phi_s^T M\phi_r \tag{3.52}$$

Comparing Equations (3.47) and (3.52), we conclude that

$$\omega_r^2 \phi_s^T M\phi_r = \omega_s^2 \phi_s^T M\phi_r \tag{3.53}$$

that is,

$$\left(\omega_r^2 - \omega_s^2\right)\phi_s^T M\phi_r = 0 \tag{3.54}$$

If the rth mode and the sth mode are distinct, then $\omega_r^2 \neq \omega_s^2$, implying that

$$\phi_s^T M\phi_r = 0 \quad \left(r \neq s\right) \tag{3.55}$$

Thus, eigenvectors ϕ_r and ϕ_s are orthogonal with respect to the mass matrix M. This holds regardless of whether M is diagonal or not, as long as it is symmetric.

Substituting result (3.55) into Equation (3.47), we obtain

$$\phi_s^T K \phi_r = 0 \quad (r \neq s) \tag{3.56}$$

showing that ϕ_r and ϕ_s are also orthogonal with respect to the stiffness matrix K.

3.10 GENERALIZED MASS AND STIFFNESS MATRICES

For $s = r$, $\phi_s^T M \phi_r$ and $\phi_s^T K \phi_r$ in Equations (3.55) and (3.56) are not equal to zero. Let

$$\phi_r^T M \phi_r = M_r \tag{3.57}$$

$$\phi_r^T K \phi_r = K_r \tag{3.58}$$

M_r and K_r are called the *generalized mass* and the *generalized stiffness* for the rth mode. Note that M_r and K_r are single values, not matrices.

For $s = r$, Equation (3.52) becomes

$$\phi_r^T K \phi_r = \omega_r^2 \phi_r^T M \phi_r \tag{3.59}$$

which, upon using expressions (3.57) and (3.58), results in the relation

$$K_r = \omega_r^2 M_r \tag{3.60}$$

From the definition of the modal matrix Φ given in Equation (3.43), we may write $\Phi^T M \Phi$ in the following expanded form:

$$\Phi^T M \Phi = \begin{bmatrix} \phi_1^T M \phi_1 & \phi_1^T M \phi_2 & . & . & \phi_1^T M \phi_n \\ \phi_2^T M \phi_1 & \phi_2^T M \phi_2 & . & . & \phi_2^T M \phi_n \\ . & . & . & . & . \\ . & . & . & . & . \\ \phi_n^T M \phi_1 & \phi_n^T M \phi_2 & . & . & \phi_n^T M \phi_n \end{bmatrix} \tag{3.61}$$

Making use of the orthogonality relation (3.55), we obtain

$$\Phi^T M \Phi = \begin{bmatrix} \phi_1^T M \phi_1 & 0 & . & . & 0 \\ 0 & \phi_2^T M \phi_2 & . & . & 0 \\ . & . & . & . & . \\ . & . & . & . & . \\ 0 & 0 & . & . & \phi_n^T M \phi_n \end{bmatrix} = \begin{bmatrix} M_1 & & & & \\ & M_2 & & & \\ & & . & & \\ & & & . & \\ & & & & M_n \end{bmatrix} = \overline{M}$$

(3.62)

Similarly, $\Phi^T K \Phi$ may be written in the expanded form

$$\Phi^T K \Phi = \begin{bmatrix} \phi_1^T K \phi_1 & \phi_1^T K \phi_2 & . & . & \phi_1^T K \phi_n \\ \phi_2^T K \phi_1 & \phi_2^T K \phi_2 & . & . & \phi_2^T K \phi_n \\ . & . & . & . & . \\ . & . & . & . & . \\ \phi_n^T K \phi_1 & \phi_n^T K \phi_2 & . & . & \phi_n^T K \phi_n \end{bmatrix}$$

(3.63)

which, upon making use of the orthogonality relation (3.56), becomes

$$\Phi^T K \Phi = \begin{bmatrix} \phi_1^T K \phi_1 & 0 & . & . & 0 \\ 0 & \phi_2^T K \phi_2 & . & . & 0 \\ . & . & . & . & . \\ . & . & . & . & . \\ 0 & 0 & . & . & \phi_n^T K \phi_n \end{bmatrix}$$

$$= \begin{bmatrix} K_1 & & & & \\ & K_2 & & & \\ & & . & & \\ & & & . & \\ & & & & K_n \end{bmatrix} = \begin{bmatrix} \omega_1^2 M_1 & & & & \\ & \omega_2^2 M_2 & & & \\ & & . & & \\ & & & . & \\ & & & & \omega_n^2 M_n \end{bmatrix} = \overline{K}$$

(3.64)

The diagonal matrices $\overline{M} = \Phi^T M \Phi$ and $\overline{K} = \Phi^T K \Phi$ (where Φ is the modal matrix, M the mass matrix and K the stiffness matrix) are called the *generalized mass matrix* and the *generalized stiffness matrix*, respectively. As earlier stated, the elements M_r and K_r ($r = 1, 2, \ldots, n$) of these matrices are the generalized mass and the generalized stiffness, respectively, for the rth mode.

3.11 WORKED EXAMPLES

Example 3.3

Obtain the generalized mass and generalized stiffness matrices for the three-storey shear frame of Example 3.1 of Section 3.7.

We have, from Equation (3.57),

$$M_1 = \phi_1^T M \phi_1 = \begin{bmatrix} 1.000 & 0.770 & 0.380 \end{bmatrix} \begin{bmatrix} m & 0 & 0 \\ 0 & 2m & 0 \\ 0 & 0 & 2m \end{bmatrix} \begin{bmatrix} 1.000 \\ 0.770 \\ 0.380 \end{bmatrix}$$

$$= \begin{bmatrix} m & 1.54m & 0.76m \end{bmatrix} \begin{bmatrix} 1.000 \\ 0.770 \\ 0.380 \end{bmatrix} = 2.4746m$$

$$M_2 = \phi_2^T M \phi_2 = \begin{bmatrix} 1.000 & -0.341 & -0.625 \end{bmatrix} \begin{bmatrix} m & 0 & 0 \\ 0 & 2m & 0 \\ 0 & 0 & 2m \end{bmatrix} \begin{bmatrix} 1.000 \\ -0.341 \\ -0.625 \end{bmatrix}$$

$$= \begin{bmatrix} m & -0.682m & -1.250m \end{bmatrix} \begin{bmatrix} 1.000 \\ -0.341 \\ -0.625 \end{bmatrix} = 2.0138m$$

$$M_3 = \phi_3^T M \phi_3 = \begin{bmatrix} 1.000 & -1.429 & 1.579 \end{bmatrix} \begin{bmatrix} m & 0 & 0 \\ 0 & 2m & 0 \\ 0 & 0 & 2m \end{bmatrix} \begin{bmatrix} 1.000 \\ -1.429 \\ 1.579 \end{bmatrix}$$

$$= \begin{bmatrix} m & -2.858m & 3.158m \end{bmatrix} \begin{bmatrix} 1.000 \\ -1.429 \\ 1.579 \end{bmatrix} = 10.0706m$$

Using the above results in Equation (3.61), and rounding off all values to three decimal places, we obtain

$$\overline{M} = \begin{bmatrix} M_1 & 0 & 0 \\ 0 & M_2 & 0 \\ 0 & 0 & M_3 \end{bmatrix} = \begin{bmatrix} 2.475m & 0 & 0 \\ 0 & 2.014m & 0 \\ 0 & 0 & 10.071m \end{bmatrix}$$

Applying Equation (3.64), with the natural circular frequencies as obtained in Example 3.1 of Section 3.7 and the generalized masses as obtained above, we have

$$
\bar{K} =
\begin{bmatrix}
K_1 & 0 & 0 \\
0 & K_2 & 0 \\
0 & 0 & K_3
\end{bmatrix}
=
\begin{bmatrix}
\omega_1^2 M_1 & 0 & 0 \\
0 & \omega_2^2 M_2 & 0 \\
0 & 0 & \omega_3^2 M_3
\end{bmatrix}
$$

$$
=
\begin{bmatrix}
\left(0.461\dfrac{k}{m}\right)2.4746m & 0 & 0 \\
0 & \left(2.682\dfrac{k}{m}\right)2.0138m & 0 \\
0 & 0 & \left(4.857\dfrac{k}{m}\right)10.0706m
\end{bmatrix}
$$

$$
=
\begin{bmatrix}
1.141k & 0 & 0 \\
0 & 5.401k & 0 \\
0 & 0 & 48.913k
\end{bmatrix}
$$

The reader may easily verify, ignoring small rounding-off discrepancies, that

$$
\Phi^T M \Phi =
\begin{bmatrix}
1.000 & 0.770 & 0.380 \\
1.000 & -0.341 & -0.625 \\
1.000 & -1.429 & 1.579
\end{bmatrix}
\begin{bmatrix}
m & 0 & 0 \\
0 & 2m & 0 \\
0 & 0 & 2m
\end{bmatrix}
$$

$$
\times
\begin{bmatrix}
1.000 & 1.000 & 1.000 \\
0.770 & -0.341 & -1.429 \\
0.380 & -0.625 & 1.579
\end{bmatrix}
= \bar{M}
$$

$$
\Phi^T K \Phi =
\begin{bmatrix}
1.000 & 0.770 & 0.380 \\
1.000 & -0.341 & -0.625 \\
1.000 & -1.429 & 1.579
\end{bmatrix}
\begin{bmatrix}
2k & -2k & 0 \\
-2k & 5k & -3k \\
0 & -3k & 7k
\end{bmatrix}
$$

$$
\times
\begin{bmatrix}
1.000 & 1.000 & 1.000 \\
0.770 & -0.341 & -1.429 \\
0.380 & -0.625 & 1.579
\end{bmatrix}
= \bar{K}
$$

However, it is always more efficient to take advantage of the orthogonality relations (3.55) and (3.56), by computing only the nonzero terms M_r and K_r (making up the diagonal matrices $\bar{M}$ and $\bar{K}$, respectively) via relations (3.57) to (3.60).

Example 3.4
Obtain the generalized mass and generalized stiffness matrices for the beam–mass system of Example 3.2 of Section 3.7.

Making use of Equation (3.57), the generalized masses are obtained as follows:

$$M_1 = \phi_1^T M \phi_1 = \begin{bmatrix} 1.000 & 1.148 \end{bmatrix} \begin{bmatrix} m & 0 \\ 0 & 3m \end{bmatrix} \begin{bmatrix} 1.000 \\ 1.148 \end{bmatrix}$$

$$= \begin{bmatrix} m & 3.444m \end{bmatrix} \begin{bmatrix} 1.000 \\ 1.148 \end{bmatrix} = 4.9537m$$

$$M_2 = \phi_2^T M \phi_2 = \begin{bmatrix} 1.000 & -0.290 \end{bmatrix} \begin{bmatrix} m & 0 \\ 0 & 3m \end{bmatrix} \begin{bmatrix} 1.000 \\ -0.290 \end{bmatrix}$$

$$= \begin{bmatrix} m & -0.870m \end{bmatrix} \begin{bmatrix} 1.000 \\ -0.290 \end{bmatrix} = 1.2523m$$

Using the above results in Equation (3.61), and rounding off all values to three decimal places, we obtain

$$\bar{M} = \begin{bmatrix} M_1 & 0 \\ 0 & M_2 \end{bmatrix} = \begin{bmatrix} 4.954m & 0 \\ 0 & 1.252m \end{bmatrix}$$

Applying Equation (3.64), using the values of natural circular frequencies obtained in Example 3.2 of Section 3.7, and the above values of the generalized masses, we obtain

$$\bar{K} = \begin{bmatrix} K_1 & 0 \\ 0 & K_2 \end{bmatrix} = \begin{bmatrix} \omega_1^2 M_1 & 0 \\ 0 & \omega_2^2 M_2 \end{bmatrix}$$

$$= \begin{bmatrix} \left(\dfrac{0.03021}{rm}\right) 4.9537m & 0 \\ 0 & \left(\dfrac{0.34479}{rm}\right) 1.2523m \end{bmatrix}$$

$$= \begin{bmatrix} \dfrac{0.150}{r} & 0 \\ 0 & \dfrac{0.432}{r} \end{bmatrix}$$

Again, the reader may easily verify that

$$\Phi^T M \Phi = \begin{bmatrix} 1.000 & 1.148 \\ 1.000 & -0.290 \end{bmatrix} \begin{bmatrix} m & 0 \\ 0 & 3m \end{bmatrix} \begin{bmatrix} 1.000 & 1.000 \\ 1.148 & -0.290 \end{bmatrix} = \bar{M}$$

$$\Phi^T K \Phi = \Phi^T D^{-1} \Phi = \begin{bmatrix} 1.000 & 1.148 \\ 1.000 & -0.290 \end{bmatrix} \begin{bmatrix} \dfrac{9}{32r} & -\dfrac{7}{32r} \\ -\dfrac{7}{32r} & \dfrac{9}{32r} \end{bmatrix}$$

$$\times \begin{bmatrix} 1.000 & 1.000 \\ 1.148 & -0.290 \end{bmatrix} = \bar{K}$$

noting that the stiffness matrix K of the system is given by the inverse of the flexibility matrix D that was derived in Example 3.2 of Section 3.7.

3.12 MODAL ANALYSIS

In order to simplify the determination of the *forced response* of multi-degree-of-freedom systems, we may decouple the differential equations of motion into independent single equations, one for each normal mode of vibration, and solve these separately as one degree-of-freedom models. The net response of the system is then simply the sum of the individual responses associated with each mode of vibration. This powerful procedure is known as *modal analysis*, or the *mode superposition method*.

For the same reasons as were given at the beginning of Section 3.5, we limit ourselves to the case of undamped vibration, for which the equation of motion for the forced response of a system with $n(n \geq 2)$ degrees of freedom is as follows:

$$M\ddot{X} + KX = F \tag{3.65}$$

where, as a reminder, the displacement vector X and the force vector F are given by

$$X = \begin{bmatrix} x_1 & x_2 & . & . & x_n \end{bmatrix}^T \tag{3.66}$$

$$F = \begin{bmatrix} f_1 & f_2 & . & . & f_n \end{bmatrix}^T \tag{3.67}$$

while the mass matrix M and stiffness matrix K are $n \times n$ symmetric square matrices, as given in earlier sections.

Let us introduce the coordinate transformation

$$X = \Phi Q \tag{3.68}$$

where Φ is the modal matrix and Q is a vector of new coordinates:

$$Q = \begin{bmatrix} q_1 & q_2 & \cdot & \cdot & q_n \end{bmatrix}^T \tag{3.69}$$

Substituting the value for X given by expression (3.68) into Equation (3.65), the equation of motion becomes

$$M\Phi\ddot{Q} + K\Phi Q = F \tag{3.70}$$

Premultiplying all terms of this equation by Φ^T, we obtain

$$\Phi^T M\Phi\ddot{Q} + \Phi^T K\Phi Q = \Phi^T F \tag{3.71}$$

The terms $\Phi^T M\Phi$ and $\Phi^T K\Phi$ are, of course, the generalized mass matrix $\overline{M}$ and generalized stiffness matrix $\overline{K}$, while the right-hand side of Equation (3.71) may be denoted by $\overline{F}$, so that we may write

$$\overline{M}\ddot{Q} + \overline{K}Q = \overline{F} \tag{3.72}$$

Writing Equation (3.72) in expanded form, we obtain

$$\begin{bmatrix} M_1 & & & \\ & M_2 & & \\ & & \ddots & \\ & & & M_n \end{bmatrix} \begin{bmatrix} \ddot{q}_1 \\ \ddot{q}_2 \\ \cdot \\ \cdot \\ \ddot{q}_n \end{bmatrix}$$

$$+ \begin{bmatrix} \omega_1^2 M_1 & & & \\ & \omega_2^2 M_2 & & \\ & & \ddots & \\ & & & \omega_n^2 M_n \end{bmatrix} \begin{bmatrix} q_1 \\ q_2 \\ \cdot \\ \cdot \\ q_n \end{bmatrix} = \begin{bmatrix} \phi_1^T F \\ \phi_2^T F \\ \cdot \\ \cdot \\ \phi_n^T F \end{bmatrix} \tag{3.73}$$

which, clearly, is a set of *decoupled* (that is, independent) equations

$$\ddot{q}_1 + \omega_1^2 q_1 = \frac{1}{M_1}\left[(X_1)_1 f_1 + (X_2)_1 f_2 + \; ... \; + (X_n)_1 f_n\right]$$

$$\ddot{q}_2 + \omega_2^2 q_2 = \frac{1}{M_2}\left[(X_1)_2 f_1 + (X_2)_2 f_2 + \; ... \; + (X_n)_2 f_n\right]$$

$$\ddot{q}_n + \omega_n^2 q_n = \frac{1}{M_n}\left[(X_1)_n f_1 + (X_2)_n f_2 + \; ... \; + (X_n)_n f_n\right] \tag{3.74}$$

in the new coordinates $\{q_1, q_2, \; ..., \; q_n\}$, which are usually referred to as *principal coordinates*. The rth equation, corresponding to parameters of the rth mode of vibration, reads

$$\ddot{q}_r + \omega_r^2 q_r = \frac{1}{M_r}\left[(X_1)_r f_1 + (X_2)_r f_2 + \; ... \; + (X_n)_r f_n\right] \tag{3.75}$$

where q_r is the principal coordinate for the rth mode; ω_r is the undamped natural circular frequency for the rth mode; M_r is the generalized mass for the rth mode; $(X_1)_r, (X_2)_r, \; ..., \; (X_n)_r$ are elements of the eigenvector for the rth mode, normalized with respect to X_1 (that is, with X_1 set equal to unity, and $X_2, X_3, \; ..., \; X_n$ assuming appropriate numerical values relative to X_1); and $f_1, f_2, \; ..., \; f_n$ are, of course, the excitation forces upon masses $m_{11}, m_{22}, \; ..., \; m_{nn}$ of the system.

Equation (3.75) is similar in form to the equation of motion for the undamped forced vibration of a one degree-of-freedom system, obtained by setting parameter c in Equation (2.1) equal to zero:

$$m\ddot{x} + kx = f(t) \tag{3.76}$$

which may be written as

$$\ddot{x} + \frac{k}{m}x = \frac{1}{m}f(t) \tag{3.77}$$

or

$$\ddot{x} + \omega^2 x = \frac{1}{m}f(t) \tag{3.78}$$

where ω is the undamped natural circular frequency of the one degree-of-freedom system. Thus, Equation (3.75) may be solved just like that of a one degree-of-freedom system, in which the coordinate x is replaced with the coordinate q_r, the parameter ω by ω_r, the mass m with the generalized mass M_r, and the forcing function $f(t)$ with the linear combination $\left[(X_1)_r f_1 + (X_2)_r f_2 + \ ... \ + (X_n)_r f_n\right]$ of system forcing functions.

For instance, if the system excitations $f_1, f_2, ..., f_n$ are all harmonic forces of the same circular frequency Ω but different amplitudes, as given by the general expression

$$f_i(t) = g_i \sin \Omega t \quad (i = 1, \ 2, \ ..., \ n) \tag{3.79}$$

where g_i is the amplitude of the force f_i at the location of the system coordinate x_i, we may write the force vector as

$$F = \begin{bmatrix} g_1 & g_2 & \cdots & g_n \end{bmatrix}^T \sin \Omega t = G \sin \Omega t \tag{3.80}$$

where $G = \begin{bmatrix} g_1 & g_2 & \cdots & g_n \end{bmatrix}^T$ is the vector of force amplitudes. The linear combination $\left[(X_1)_r f_1 + (X_2)_r f_2 + \ ... \ + (X_n)_r f_n\right]$ for the rth mode may therefore be written as

$$(X_1)_r f_1 + (X_2)_r f_2 + \ ... \ + (X_n)_r f_n = \left[\phi_r^T G\right]\sin \Omega t = G_r \sin \Omega t \tag{3.81}$$

where $G_r = \phi_r^T G$ is a constant (the *generalized amplitude*) for the rth mode. In this case, Equation (3.75) becomes

$$\ddot{q}_r + \omega_r^2 q_r = \frac{1}{M_r} G_r \sin \Omega t \tag{3.82}$$

Now, for a single degree-of-freedom system subjected to a harmonic excitation of the form given by Equation (2.80), rewritten here as

$$\ddot{x} + \omega^2 x = \frac{1}{m} g \sin \Omega t \tag{3.83}$$

for ease of comparisons, the solution for undamped motion was given in Chapter 2 as Equation (2.90). Noting the similarity of Equations (3.82) and (3.83), we may write down the general solution of Equation (3.82) directly from that of Equation (3.83), as follows:

$$q_r = \left(\frac{G_r / K_r}{1 - r_r^2}\right)\sin \Omega t + A_{1r} \sin \omega_r t + A_{2r} \cos \omega_r t \tag{3.84}$$

where $K_r \left(= \omega_r^2 M_r \right)$ is the generalized stiffness for the rth mode (which is analogous to $k = \omega^2 m$ for a single degree-of-freedom system), and $r_r (= \Omega / \omega_r)$ is the frequency ratio for the rth mode; the constants A_{1r} and A_{2r} for the rth mode can only be determined simultaneously with those of other modes, when initial conditions are applied upon real displacements x_i ($i = 1, 2, ..., n$) and real velocities $\dot{x}_i$ ($i = 1, 2, ..., n$) of the system, as will be elaborated shortly.

Having obtained general solutions to Equation (3.75) for all the q_r's ($r = 1, 2, ..., n$) of a problem, irrespective of the type of excitation functions that are applied to the system, the actual system displacements follow from Equation (3.68), that is,

$$
\begin{bmatrix} x_1 \\ x_2 \\ . \\ . \\ x_n \end{bmatrix} = \begin{bmatrix} X_1 \\ X_2 \\ . \\ . \\ X_n \end{bmatrix}_1 \begin{bmatrix} X_1 \\ X_2 \\ . \\ . \\ X_n \end{bmatrix}_2 \quad . \quad . \quad \begin{bmatrix} X_1 \\ X_2 \\ . \\ . \\ X_n \end{bmatrix}_n \begin{bmatrix} q_1 \\ q_2 \\ . \\ . \\ q_n \end{bmatrix} \tag{3.85}
$$

which, upon writing the equations of this system separately, yields

$$
x_1 = \left(X_1 \right)_1 q_1 + \left(X_1 \right)_2 q_2 + ... + \left(X_1 \right)_n q_n
$$

$$
x_2 = \left(X_2 \right)_1 q_1 + \left(X_2 \right)_2 q_2 + ... + \left(X_2 \right)_n q_n
$$

$$
.
$$

$$
.
$$

$$
x_n = \left(X_n \right)_1 q_1 + \left(X_n \right)_2 q_2 + ... + \left(X_n \right)_n q_n \tag{3.86}
$$

where $(X_i)_j$ is, of course, the ith value of the eigenvector for mode j.

From the $2n$ initial conditions

$$
(x_i)_{t=0} = x_{i0} \quad (i = 1, 2, ..., n) \tag{3.87}
$$

$$
(\dot{x}_i)_{t=0} = \dot{x}_{i0} \quad (i = 1, 2, ..., n) \tag{3.88}
$$

for displacement and velocity of each of the n masses of the system, we can form $2n$ simultaneous equations in $2n$ unknowns $\left(\{ A_{1r}, A_{2r} \}, \ r = 1, 2, ..., n \right)$. In the case of synchronous harmonic excitation considered earlier, these unknowns are the constants $\{A_{1r}, A_{2r}\}$ introduced in the general solution for q_r (Equation (3.84)).

The $2n$ simultaneous equations may then be solved for the constants ($\{A_{1r}, A_{2r}\}, r = 1, 2, ..., n$), permitting the displacements to be fully known.

3.13 WORKED EXAMPLE

Example 3.5
For the three-storey shear frame of Examples 3.1 and 3.3, sychronous harmonic excitations of circular frequency Ω act on the three slab masses, as shown in Figure 3.12. Derive expressions for the response $\{x_1(t), x_2(t), x_3(t)\}$ of the system.

The decoupled equations of motion in principal coordinates are

$$\ddot{q}_1 + \omega_1^2 q_1 = \frac{1}{M_1}\left(\phi_1^T F\right)$$

$$\ddot{q}_2 + \omega_2^2 q_2 = \frac{1}{M_2}\left(\phi_2^T F\right)$$

$$\ddot{q}_3 + \omega_3^2 q_3 = \frac{1}{M_3}\left(\phi_3^T F\right)$$

The force vector is F given by

$$F = \begin{bmatrix} 3H\sin\Omega t \\ 2H\sin\Omega t \\ H\sin\Omega t \end{bmatrix} = \begin{bmatrix} 3H \\ 2H \\ H \end{bmatrix}\sin\Omega t$$

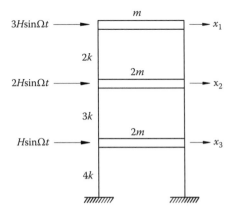

Figure 3.12 Harmonically excited three-storey shear frame of Example 3.5.

The eigenvectors $\{\phi_1, \phi_2, \phi_3\}$ were obtained in Example 3.1. Using these, and the above vector F, we obtain

$$\phi_1^T F = \begin{bmatrix} 1.000 & 0.770 & 0.380 \end{bmatrix} \begin{bmatrix} 3H \\ 2H \\ H \end{bmatrix} \sin\Omega t$$

$$= (4.920H)\sin\Omega t = G_1 \sin\Omega t$$

$$\phi_2^T F = \begin{bmatrix} 1.000 & -0.341 & -0.625 \end{bmatrix} \begin{bmatrix} 3H \\ 2H \\ H \end{bmatrix} \sin\Omega t$$

$$= (1.693H)\sin\Omega t = G_2 \sin\Omega t$$

$$\phi_3^T F = \begin{bmatrix} 1.000 & -1.429 & 1.579 \end{bmatrix} \begin{bmatrix} 3H \\ 2H \\ H \end{bmatrix} \sin\Omega t$$

$$= (1.721H)\sin\Omega t = G_3 \sin\Omega t$$

Applying the general solution for q_r (Equation (3.84)), we have

$$q_1 = \frac{G_1/K_1}{1-r_1^2} \sin\Omega t + A_{11}\sin\omega_1 t + A_{21}\cos\omega_1 t$$

$$q_2 = \frac{G_2/K_2}{1-r_2^2} \sin\Omega t + A_{12}\sin\omega_2 t + A_{22}\cos\omega_2 t$$

$$q_3 = \frac{G_3/K_3}{1-r_3^2} \sin\Omega t + A_{13}\sin\omega_3 t + A_{23}\cos\omega_3 t$$

where, as found in Example 3.1, the natural circular frequencies are

$$\omega_1 = 0.679\sqrt{\frac{k}{m}} \quad \text{for mode 1}$$

$$\omega_2 = 1.638\sqrt{\frac{k}{m}} \quad \text{for mode 2}$$

$$\omega_3 = 2.204\sqrt{\frac{k}{m}} \quad \text{for mode 3}$$

so that the squares of the frequency ratios become

$$r_1^2 = \left(\frac{\Omega}{\omega_1}\right)^2 = \frac{m\Omega^2}{0.461k} \quad \text{for mode 1}$$

$$r_2^2 = \left(\frac{\Omega}{\omega_2}\right)^2 = \frac{m\Omega^2}{2.683k} \quad \text{for mode 2}$$

$$r_3^2 = \left(\frac{\Omega}{\omega_3}\right)^2 = \frac{m\Omega^2}{4.857k} \quad \text{for mode 3}$$

From the earlier evaluations of the parameters $\phi_i^T F$ $(i = 1, 2, 3)$, the generalized amplitudes were obtained as

$G_1 = 4.920H$ for mode 1

$G_2 = 1.693H$ for mode 2

$G_3 = 1.721H$ for mode 3

The generalized stiffnesses were obtained in Example 3.3 as follows:

$K_1 = 1.141k$ for mode 1

$K_2 = 5.401k$ for mode 2

$K_3 = 48.913k$ for mode 3

Upon substituting the above values in the earlier expressions for the generalized coordinates, we obtain

$$q_1 = \left(\frac{1.988H}{0.461k - m\Omega^2}\right)\sin\Omega t + A_{11}\sin\omega_1 t + A_{21}\cos\omega_1 t$$

$$q_2 = \left(\frac{0.841H}{2.683k - m\Omega^2}\right)\sin\Omega t + A_{12}\sin\omega_2 t + A_{22}\cos\omega_2 t$$

$$q_3 = \left(\frac{0.171H}{4.857k - m\Omega^2}\right)\sin\Omega t + A_{13}\sin\omega_3 t + A_{23}\cos\omega_3 t$$

Transforming back to ordinary coordinates, we have

$$x_1 = (X_1)_1 q_1 + (X_1)_2 q_2 + (X_1)_3 q_3$$

$$= 1.000 q_1 + 1.000 q_2 + 1.000 q_3$$

$$= 1.000 \left[\left(\frac{1.988H}{0.461k - m\Omega^2} \right) \sin \Omega t + A_{11} \sin \omega_1 t + A_{21} \cos \omega_1 t \right]$$

$$+ 1.000 \left[\left(\frac{0.841H}{2.683k - m\Omega^2} \right) \sin \Omega t + A_{12} \sin \omega_2 t + A_{22} \cos \omega_2 t \right]$$

$$+ 1.000 \left[\left(\frac{0.171H}{4.857k - m\Omega^2} \right) \sin \Omega t + A_{13} \sin \omega_3 t + A_{23} \cos \omega_3 t \right]$$

$$x_2 = (X_2)_1 q_1 + (X_2)_2 q_2 + (X_2)_3 q_3$$

$$= 0.770 q_1 - 0.341 q_2 - 1.429 q_3$$

$$= 0.770 \left[\left(\frac{1.988H}{0.461k - m\Omega^2} \right) \sin \Omega t + A_{11} \sin \omega_1 t + A_{21} \cos \omega_1 t \right]$$

$$- 0.341 \left[\left(\frac{0.841H}{2.683k - m\Omega^2} \right) \sin \Omega t + A_{12} \sin \omega_2 t + A_{22} \cos \omega_2 t \right]$$

$$- 1.429 \left[\left(\frac{0.171H}{4.857k - m\Omega^2} \right) \sin \Omega t + A_{13} \sin \omega_3 t + A_{23} \cos \omega_3 t \right]$$

$$x_3 = (X_3)_1 q_1 + (X_3)_2 q_2 + (X_3)_3 q_3$$

$$= 0.380 q_1 - 0.625 q_2 + 1.579 q_3$$

$$= 0.380 \left[\left(\frac{1.988H}{0.461k - m\Omega^2} \right) \sin \Omega t + A_{11} \sin \omega_1 t + A_{21} \cos \omega_1 t \right]$$

$$- 0.625 \left[\left(\frac{0.841H}{2.683k - m\Omega^2} \right) \sin \Omega t + A_{12} \sin \omega_2 t + A_{22} \cos \omega_2 t \right]$$

$$+ 1.579 \left[\left(\frac{0.171H}{4.857k - m\Omega^2} \right) \sin \Omega t + A_{13} \sin \omega_3 t + A_{23} \cos \omega_3 t \right]$$

where the six constants $\{A_{11}, A_{21}, A_{12}, A_{22}, A_{13}, A_{23}\}$ are to be obtained by specifying the six initial conditions $\{x_1(0), \dot{x}_1(0), x_2(0), \dot{x}_2(0), x_3(0), \dot{x}_3(0)\}$.

TUTORIAL QUESTIONS

Unless otherwise stated, assume damping is negligible in all problems.

3.1. A vertical cantilever structure is modelled as two concentrated masses $2m$ and m, at heights h and $2h$, respectively, above the ground, and supported on a light column whose lower portion (between ground level and the lower mass) is of flexural rigidity $3EI$ and upper portion (between the two masses) is of flexural rigidity EI (Figure Q3.1). Obtain the natural frequencies and mode shapes of the system, for small lateral motions x_1 and x_2 of the upper and lower masses, respectively.

3.2. A beam of total length $4l$ and constant flexural rigidity EI is simply supported at two points equidistant from its middle point, such that there is an overhang of length l at each end (Figure Q3.2). Find the natural frequencies and mode shapes for the small transverse vibrations of the beam, when it is modelled as concentrated masses m, $2m$ and m, at the midpoints of the left overhang, the supported span, and the right overhang, respectively.

3.3. Four equal masses, numbered as shown in Figure Q3.3, are interconnected by springs, such that the entire system lies in one plane and

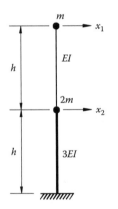

Figure Q3.1 Vertical cantilever structure of Question 3.1.

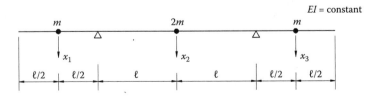

Figure Q3.2 Horizontal beam of Question 3.2.

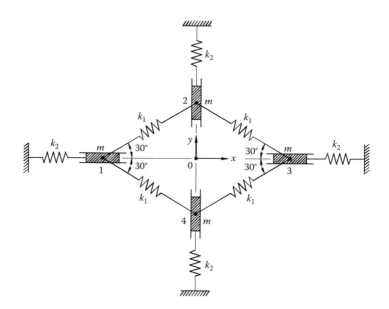

Figure Q3.3 Spring–mass system of Question 3.3.

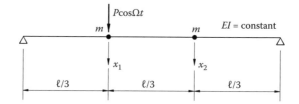

Figure Q3.4 Harmonically loaded beam of Question 3.4.

has a configuration that is symmetrical about the x and y Cartesian axes, whose origin O is at the centre of the configuration. By means of frictionless guide channels, the masses are constrained to move either only in the direction coinciding with the x axis or only in the direction coinciding with the y axis, as shown in the figure. Assuming deformations of the springs are small, obtain the stiffness matrix of the system.

3.4. Obtain the undamped forced response of the beam–mass model shown in Figure Q3.4. The beam is simply supported, of length l, and modelled as two concentrated masses each of magnitude m, lumped at locations a third of the span from either end. A harmonic excitation $P\cos\Omega t$ acts vertically upon the left mass. Assume the initial conditions are $x_1(0) = \dot{x}_1(0) = x_2(0) = \dot{x}_2(0) = 0$.

Chapter 4

Continuous systems

4.1 INTRODUCTION

So far, we have dealt with the vibration of structures modelled as concentrated masses. Real structures have distributed mass, and for better accuracy, it is often desirable to model these as continuous systems. Modelling a structure as a discrete system (that is, a lumped parameter model) involves a finite number of degrees of freedom, whereas modelling the same structure as a continuous system involves an infinite number of degrees of freedom.

A continuous system may be visualized as a discrete system in which the number of degrees of freedom tends to infinity. In general, the mathematical modelling of discrete systems results in *ordinary* differential equations involving time as the only independent variable, while that of continuous systems results in *partial* differential equations involving both time and space as the independent variables.

In this chapter, we will consider the mathematical modelling of the vibration behaviour of a number of structural systems as continuous systems, and obtain closed-form solutions for these. Systems amenable to such solution procedures include the transverse vibration of strings, the axial vibration of rods, the torsional vibration of shafts and the flexural vibration of beams and plates. Where the ensuing differential equations are difficult or impossible to solve analytically, we have to resort to numerical methods such as the finite-element method (Chapter 5) or the finite-difference method.

In all the considerations that follow, we will assume the displacements during vibration are small in comparison with the relevant dimension of the system (such as the length of a string or a rod, the depth of a beam or the thickness of a plate), so that the elastic response remains essentially linear throughout. Formulations for static analyses also make the same assumption.

4.2 TRANSVERSE VIBRATION OF STRINGS

4.2.1 General formulation

Figure 4.1(a) shows a displaced string of length l and mass per unit length $m(x)$. Although strings usually have a constant mass per unit length, our formulation will allow variation of the mass per unit length with respect to the coordinate x along the string. Let the tension in the string be $T(x)$, which may also vary with the coordinate x. The string is undergoing small transverse vibrations $v(x,t)$ in the y direction, under the excitation of a transverse force per unit length $f(x,t)$, also in the y direction. Note that both the transverse displacement and the intensity of the excitation force are functions of the space coordinate x and the time t.

In Figure 4.1(b), the forces acting on an element dx of the string are shown as a free-body diagram. Both the tension T and the slope of the string dv/dx are shown incremented on the positive side of the element, by amounts equal to the rate of variation of the quantity with respect to the coordinate x, multiplied by the element length dx. Since the intensity of the applied external force is f per unit length and the mass of the string is m per unit length, the resultant excitation force on the element is $f\,dx$, while the resultant inertial

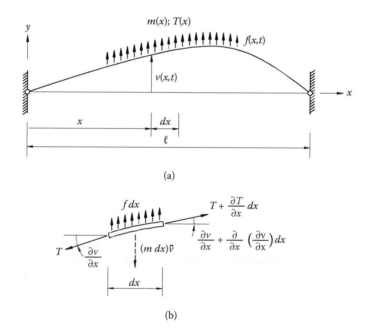

(a)

(b)

Figure 4.1 Transverse vibration of a string: (a) displaced profile, (b) forces acting on an element dx of the string.

force is $(m\,dx)\ddot{v}$, which represents the total mass of the element $m\,dx$ multiplied by its acceleration $\ddot{v}$. As in previous chapters, the notation $\dot{v}$ and $\ddot{v}$ is used to denote the first and second derivatives, respectively, of the displacement v with respect to time, which, of course, are the velocity and the acceleration.

Resolving the forces in Figure 4.1(b) in the positive y direction, we obtain

$$
\left[T(x) + \frac{\partial T(x)}{\partial x}dx\right]\left[\frac{\partial v(x, t)}{\partial x} + \frac{\partial^2 v(x, t)}{\partial x^2}dx\right] - T(x)\frac{\partial v(x, t)}{\partial x} + f(x, t)\,dx
$$
$$
= m(x)\,dx\,\frac{\partial^2 v(x, t)}{\partial t^2}
$$

(4.1)

Ignoring the products of small quantities in relation to the quantities themselves (the products of small quantities are usually referred to as second-order terms), we obtain

$$
T(x)\frac{\partial^2 v(x, t)}{\partial x^2}dx + \frac{\partial T(x)}{\partial x}\frac{\partial v(x, t)}{\partial x}dx + f(x, t)\,dx = m(x)\frac{\partial^2 v(x, t)}{\partial t^2}dx \quad (4.2)
$$

Cancelling the dx from all terms, and applying the product rule of derivatives to combine the first two terms, we finally obtain

$$
\frac{\partial}{\partial x}\left[T(x)\frac{\partial v(x, t)}{\partial x}\right] + f(x, t) = m(x)\frac{\partial^2 v(x, t)}{\partial t^2}
$$

(4.3)

The above is the governing equation of motion for the transverse vibration of a tensioned string under a time-varying excitation force.

4.2.2 Free vibration

When there is no excitation force (the case of free vibration), we have $f(x, t) = 0$, so that Equation (4.3) reduces to

$$
\frac{\partial}{\partial x}\left[T(x)\frac{\partial v(x, t)}{\partial x}\right] = m(x)\frac{\partial^2 v(x, t)}{\partial t^2}
$$

(4.4)

Let us assume that the shape of the vibrating string remains the same during vibration, with all points on the string moving in synchrony, passing through the zero-displacement position at the same time, and reaching the maximum amplitude at the same time. Such motion is called *synchronous*. We may write the solution in the form

$$
v(x, t) = V(x)\,F(t)
$$

(4.5)

where $V(x)$ represents the shape of the string at any given time (a snapshot of the vibrating string), and $F(t)$ describes how the displacement at any given point on the string varies with time.

Substituting Equation (4.5) into Equation (4.4), the left-hand side of Equation (4.4) becomes

$$\frac{\partial}{\partial x}\left[T(x)\frac{\partial v(x, t)}{\partial x}\right] = \frac{\partial}{\partial x}\left[T(x)\,F(t)\,\frac{dV(x)}{dx}\right] = F(t)\,\frac{d}{dx}\left[T(x)\,\frac{dV(x)}{dx}\right] \quad (4.6a)$$

while the right-hand side of Equation (4.4) becomes

$$m(x)\,\frac{\partial^2 v(x, t)}{\partial t^2} = m(x)\,V(x)\,\frac{d^2 F(t)}{dt^2} \quad (4.6b)$$

Thus, Equation (4.4) as a whole becomes

$$F(t)\,\frac{d}{dx}\left[T(x)\,\frac{dV(x)}{dx}\right] = m(x)\,V(x)\,\frac{d^2 F(t)}{dt^2}$$

that is,

$$\frac{1}{m(x)\,V(x)}\,\frac{d}{dx}\left[T(x)\,\frac{dV(x)}{dx}\right] = \frac{1}{F(t)}\,\frac{d^2 F(t)}{dt^2} \quad (4.7)$$

Since the left-hand side of Equation (4.7) is a function of x only, and the right-hand side is a function of t only, the only way this equation can be satisfied for all x and t is if both sides are constant. We may therefore write

$$\frac{1}{F(t)}\,\frac{d^2 F(t)}{dt^2} = k \quad \Rightarrow \quad \frac{d^2 F(t)}{dt^2} = kF(t) \quad (4.8a)$$

$$\frac{1}{m(x)\,V(x)}\,\frac{d}{dx}\left[T(x)\,\frac{dV(x)}{dx}\right] = k \quad \Rightarrow \quad \frac{d}{dx}\left[T(x)\,\frac{dV(x)}{dx}\right] = k\,m(x)\,V(x) \quad (4.8b)$$

If we let the constant k equal $-\omega^2$, these equations assume the form

$$\frac{d^2 F(t)}{dt^2} + \omega^2 F(t) = 0 \quad (4.9a)$$

$$\frac{d}{dx}\left[T(x)\,\frac{dV(x)}{dx}\right] + \omega^2 m(x)\,V(x) = 0 \quad (4.9b)$$

For harmonic motion, and as seen in Chapter 2, the solution of Equation (4.9a) may be written in the form

$$F(t) = A\cos(\omega t - \alpha) \tag{4.10}$$

where ω is the circular frequency of the harmonic motion, and α is the phase angle. The constants A and α may readily be chosen to satisfy the initial conditions at $t = 0$.

To determine the shape function $V(x)$, we must solve Equation (4.9b). Let us assume the tension T in the string is constant and the mass per unit length m is also constant, which is the case in many situations encountered in practice. Equation (4.9b) then becomes

$$T\frac{d^2V(x)}{dx^2} + \omega^2 m\, V(x) = 0 \tag{4.11}$$

which we may write as

$$\frac{d^2V(x)}{dx^2} + \beta^2\, V(x) = 0 \tag{4.12}$$

where

$$\beta^2 = \frac{m\omega^2}{T} \tag{4.13}$$

Let us write the solution of Equation (4.12) as follows:

$$V(x) = C_1 \sin \beta x + C_2 \cos \beta x \tag{4.14}$$

where C_1 and C_2 are constants of integration. At the ends of the string, the displacements are zero. The condition $V(0) = 0$ gives $C_2 = 0$, so that Equation (4.14) reduces to

$$V(x) = C_1 \sin \beta x \tag{4.15}$$

Applying the condition $V(l) = 0$ to Equation (4.15), we obtain

$$C_1 \sin \beta l = 0 \tag{4.16}$$

Now C_1 cannot be zero since this would make $V(x)$ equal to zero throughout (see Equation (4.15)), implying no vibrations. Hence,

$$\sin \beta l = 0 \tag{4.17}$$

giving the solutions

$$\beta l = \pi, 2\pi, 3\pi, \ldots \tag{4.18}$$

There are an infinite number of modes corresponding to these solutions. For the nth mode, we may write

$$\beta_n l = n\pi \tag{4.19}$$

which gives

$$\beta_n = \frac{n\pi}{l} \tag{4.20}$$

For this mode, Equation (4.15) becomes

$$V_n(x) = C_{1n} \sin\frac{n\pi x}{l} \tag{4.21}$$

The amplitude C_{1n} cannot be determined. We can only know the shape of the mode. The shapes of the first three modes $V_1(x)$, $V_2(x)$ and $V_3(x)$ are depicted in Figure 4.2. From Equations (4.13) and (4.20), the circular frequency for the nth mode is given by

$$\omega_n = \frac{n\pi}{l}\sqrt{\frac{T}{m}} = n\pi\sqrt{\frac{T}{ml^2}} \tag{4.22}$$

The lowest frequency, ω_1, is called the fundamental frequency, and this corresponds to the mode shape in Figure 4.2(a) (one wave). In a stringed instrument (such as a guitar), one can vary the frequency of a string by adjusting the tension T (tuning), or modifying the effective length l of the string (by altering the position at which the finger is pressing down the string).

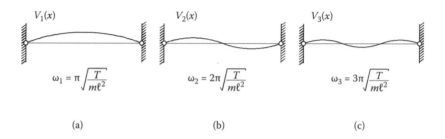

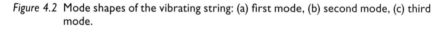

Figure 4.2 Mode shapes of the vibrating string: (a) first mode, (b) second mode, (c) third mode.

4.3 AXIAL VIBRATION OF RODS

Figure 4.3(a) shows a rod of length l experiencing axial vibrations $u(x,t)$. The mass per unit length at coordinate x is denoted by $m(x)$, while the axial rigidity is denoted by $EA(x)$, where E is the Young's modulus and A the cross-sectional area. We will assume that the rod is not subject to external excitation, and is undergoing free vibration.

Figure 4.3(b) depicts an element dx of the rod. The axial force in the rod is denoted by N, and this varies across the element as shown. The inertial force on the element is $(m\,dx)\ddot{u}$, where $\ddot{u}$ is the acceleration. Summing forces in the positive x direction,

$$N(x) + \frac{\partial N(x)}{\partial x}\,dx - N(x) - m(x)\,dx\,\frac{\partial^2 u(x,\,t)}{\partial t^2} = 0 \qquad (4.23)$$

that is,

$$\frac{\partial N(x)}{\partial x} = m(x)\,\frac{\partial^2 u(x,\,t)}{\partial t^2} \qquad (4.24)$$

Now, from Hooke's law,

$$N(x) = \sigma(x)\,A(x) = \varepsilon(x)\,EA(x) = \frac{\partial u(x)}{\partial x}\,EA(x) \qquad (4.25)$$

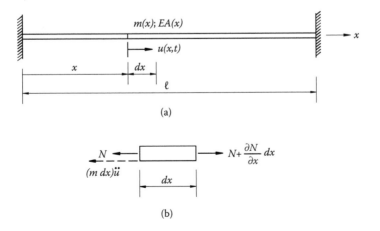

(a)

(b)

Figure 4.3 Axial vibration of a rod: (a) general configuration, (b) forces acting on an element dx of the rod.

where σ denotes axial stress and ε denotes axial strain, the latter equal to $\partial u/\partial x$ by definition. Using this expression to eliminate $N(x)$ from Equation (4.24), we obtain

$$\frac{\partial}{\partial x}\left[EA(x)\frac{\partial u(x,t)}{\partial x}\right] = m(x)\frac{\partial^2 u(x,t)}{\partial t^2} \tag{4.26}$$

Assuming synchronous motion as before, we may express $u(x,t)$ as a product of a shape function $U(x)$ and a harmonic function $F(t)$:

$$u(x,t) = U(x)F(t) \tag{4.27}$$

Substituting this expression for $u(x,t)$ into Equation (4.26), we obtain

$$\frac{\partial}{\partial x}\left[EA(x)F(t)\frac{dU(x)}{dx}\right] = m(x)U(x)\frac{d^2F(t)}{dt^2} \tag{4.28a}$$

that is,

$$F(t)\frac{d}{dx}\left[EA(x)\frac{dU(x)}{dx}\right] = m(x)U(x)\frac{d^2F(t)}{dt^2} \tag{4.28b}$$

Rearranging,

$$\frac{1}{m(x)U(x)}\frac{d}{dx}\left[EA(x)\frac{dU(x)}{dx}\right] = \frac{1}{F(t)}\frac{d^2F(t)}{dt^2} \tag{4.28c}$$

The left-hand side of Equation (4.28c) is a function of x only, while the right-hand side is a function of t only. Both sides have to be constant if this equation is to be satisfied for all x and t. If we let this constant be $-\omega^2$ as before, we may write the right-hand side as

$$\frac{d^2F(t)}{dt^2} + \omega^2 F(t) = 0 \tag{4.29}$$

which has the general solution

$$F(t) = A\cos(\omega t - \alpha) \tag{4.30}$$

The left-hand side may be written

$$\frac{d}{dx}\left[EA(x)\frac{dU(x)}{dx}\right] + \omega^2 m(x)U(x) = 0 \tag{4.31}$$

Let us assume the rod is uniform (m is constant, EA is constant). Equation (4.31) simplifies to

$$EA\frac{d^2U(x)}{dx^2} + m\omega^2 U(x) = 0 \qquad (4.32)$$

that is,

$$\frac{d^2U(x)}{dx^2} + \beta^2\, U(x) = 0 \qquad (4.33)$$

where

$$\beta^2 = \frac{m\omega^2}{EA} \qquad (4.34)$$

Equation (4.33) is exactly of the same form as Equation (4.12) for the string. If the ends of the bar are fixed, the two kinematic conditions $U(0) = 0$ and $U(l) = 0$ apply, so that the same solutions as for the string (Equation (4.20)) hold. From Equation (4.34) in conjunction with Equation (4.20), the circular natural frequencies ω_n are therefore

$$\omega_n = \beta_n\sqrt{\frac{EA}{m}} = \frac{n\pi}{l}\sqrt{\frac{EA}{m}} = n\pi\sqrt{\frac{EA}{ml^2}} \qquad (4.35)$$

The first three modes of the bar, $U_1(x)$, $U_2(x)$ and $U_3(x)$, are shown in Figure 4.4. The nonzero slopes of the plots at the fixed ends of the bar might seem inconsistent with the boundary conditions at a quick glance, but there is no contradiction, bearing in mind that we are plotting *axial* displacements, not transverse displacements.

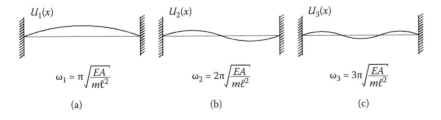

$$\omega_1 = \pi\sqrt{\frac{EA}{ml^2}} \qquad \omega_2 = 2\pi\sqrt{\frac{EA}{ml^2}} \qquad \omega_3 = 3\pi\sqrt{\frac{EA}{ml^2}}$$

(a)   (b)   (c)

Figure 4.4 Mode shapes of the vibrating rod: (a) first mode, (b) second mode, (c) third mode.

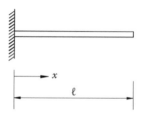

Figure 4.5 Rod fixed at one end and free at the other end.

If the rod is fixed at one end and free at the other, we have a cantilever (Figure 4.5). The relevant boundary conditions are now a kinematic and a static one:

$$U(0) = 0 \tag{4.36a}$$

$$N(l) = 0 \tag{4.36b}$$

These express the fact that the axial displacement is zero at $x = 0$, and the axial force is zero at $x = l$.

If we write the general solution of Equation (4.33) as

$$U(x) = C_1 \sin \beta x + C_2 \cos \beta x \tag{4.37}$$

(which, of course, is of the same form as Equation (4.14) for the string), we may apply the first boundary condition to obtain $C_2 = 0$. Equation (4.37) thus reduces to

$$U(x) = C_1 \sin \beta x \tag{4.38}$$

Using Equations (4.25) and (4.27), we may express the axial force N in terms of the parameters of Equation (4.38) as follows:

$$N(x) = EA \frac{\partial u}{\partial x} = EAF(t) \frac{dU(x)}{dx} = EAF(t) C_1 \beta \cos \beta x \tag{4.39}$$

Applying the second boundary condition (Equation (4.36b)), we obtain

$$\cos \beta l = 0 \tag{4.40}$$

with solutions

$$\beta l = \frac{\pi}{2}, \frac{3\pi}{2}, \frac{5\pi}{2}, \dots \tag{4.41}$$

For the nth mode, we may write

$$\beta_n l = \left(\frac{2n-1}{2}\right)\pi \tag{4.42}$$

which gives

$$\beta_n = \left(\frac{2n-1}{2l}\right)\pi \tag{4.43}$$

For this mode, Equation (4.38) becomes

$$U_n(x) = C_{1n} \sin\left(\frac{2n-1}{2l}\right)\pi x \tag{4.44}$$

The first three modes are

$$U_1(x) = C_{11} \sin\frac{\pi x}{2l} \tag{4.45a}$$

$$U_2(x) = C_{12} \sin\frac{3\pi x}{2l} \tag{4.45b}$$

$$U_3(x) = C_{13} \sin\frac{5\pi x}{2l} \tag{4.45c}$$

These are plotted in Figure 4.6. From Equations (4.34) and (4.43), the circular natural frequencies for the cantilever rod are given by

$$\omega_n = \left(\frac{2n-1}{2l}\right)\pi\sqrt{\frac{EA}{m}} = \frac{\pi}{2}(2n-1)\sqrt{\frac{EA}{ml^2}} \tag{4.46}$$

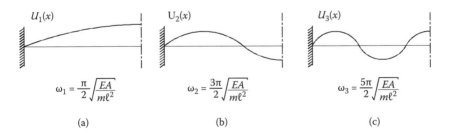

Figure 4.6 Mode shapes of the cantilever rod: (a) first mode, (b) second mode, (c) third mode.

Comparing Equations (4.46) and (4.35), we note that the fundamental frequency (corresponding to $n = 1$) for axial vibrations of the cantilever rod is only 50% of that of the rod fixed at both ends, but as the mode number (n) increases, the corresponding frequencies of the cantilever rod and the rod fixed at both ends tend towards each other.

4.4 FLEXURAL VIBRATION OF BEAMS

4.4.1 General formulation

Figure 4.7(a) shows a beam of mass per unit length $m(x)$ and flexural rigidity $EI(x)$, where $I(x)$ is the second moment of area of the beam cross section at the coordinate x. The beam is experiencing small transverse vibrations $v(x, t)$ under a transverse excitation force $f(x, t)$ per unit length. The forces acting on an element dx of the beam are shown in Figure 4.7(b). In the diagram, $M(x, t)$ and $Q(x, t)$ denote bending moment and shear force, respectively. Resolving forces in the y direction, we obtain

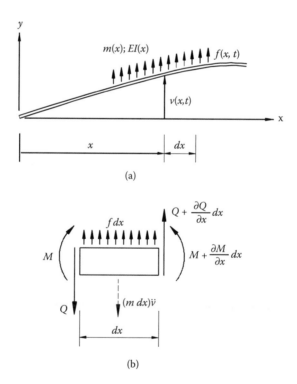

(a)

(b)

Figure 4.7 Flexural vibration of a beam: (a) displaced profile of part of the beam, (b) forces and moments acting on an element dx of the beam.

$$Q(x, t) + \frac{\partial Q(x, t)}{\partial x} dx - Q(x, t) + f(x, t) \, dx = m(x) \, dx \, \frac{\partial^2 v(x, t)}{\partial t^2} \qquad (4.47)$$

which simplifies to

$$\frac{\partial Q(x, t)}{\partial x} + f(x, t) = m(x) \frac{\partial^2 v(x, t)}{\partial t^2} \qquad (4.48)$$

Taking moments about an axis passing through the centre O of the element and perpendicular to the plane of the diagram, we obtain

$$\left[M(x, t) + \frac{\partial M(x, t)}{\partial x} dx \right] - M(x, t) + \left[Q(x, t) + \frac{\partial Q(x, t)}{\partial x} dx \right] \frac{dx}{2}$$
$$+ Q(x, t) \frac{dx}{2} = 0 \qquad (4.49)$$

Simplifying the above equation and ignoring second-order terms, we obtain

$$\frac{\partial M(x, t)}{\partial x} + Q(x, t) = 0 \qquad (4.50)$$

so that

$$Q(x, t) = - \frac{\partial M(x, t)}{\partial x} \qquad (4.51)$$

Substituting this result into Equation (4.48), we obtain

$$- \frac{\partial^2 M(x, t)}{\partial x^2} + f(x, t) = m(x) \frac{\partial^2 v(x, t)}{\partial t^2} \qquad (4.52)$$

Now, from elementary mechanics of structures,

$$M(x) = EI \frac{d^2 v(x)}{dx^2} \qquad (4.53)$$

Using this expression for the bending moment, Equation (4.52) finally becomes

$$- \frac{\partial^2}{\partial x^2} \left[EI(x) \frac{\partial^2 v(x, t)}{\partial x^2} \right] + f(x, t) = m(x) \frac{\partial^2 v(x, t)}{\partial t^2} \qquad (4.54)$$

4.4.2 Boundary conditions

If an end $x = \xi$ (where $\xi = 0$ or l) is clamped/fixed, it implies both the displacement and the slope at that end are zero (two kinematic conditions):

$$v(\xi, t) = 0; \quad \left.\frac{\partial v(x, t)}{\partial x}\right|_{x=\xi} = 0 \qquad (4.55a, b)$$

If an end $x = \xi$ is hinged/pinned, it implies both the displacement and the bending moment at that end are zero (one kinematic and one static condition):

$$v(\xi, t) = 0; \quad \left. EI(x)\frac{\partial^2 v(x, t)}{\partial x^2}\right|_{x=\xi} = 0 \qquad (4.56a, b)$$

If an end $x = \xi$ is free, it implies both the bending moment and the shear force at that end are zero (two static conditions):

$$EI(x)\left.\frac{\partial^2 v(x, t)}{\partial x^2}\right|_{x=\xi} = 0; \quad \frac{\partial M(x, t)}{\partial x} = \frac{\partial}{\partial x}\left[EI(x)\frac{\partial^2 v(x, t)}{\partial x^2}\right]_{x=\xi} = 0 \quad (4.57a, b)$$

4.4.3 Free vibration

Setting $f(x, t)$ to zero in Equation (4.54), we obtain the equation of motion for free vibration:

$$-\frac{\partial^2}{\partial x^2}\left[EI(x)\frac{\partial^2 v(x, t)}{\partial x^2}\right] = m(x)\frac{\partial^2 v(x, t)}{\partial t^2} \qquad (4.58)$$

Assuming synchronous motion as before, we may write

$$v(x, t) = V(x)\, F(t) \qquad (4.59)$$

Substituting this expression into Equation (4.58), we obtain

$$-F(t)\frac{d^2}{dx^2}\left[EI(x)\frac{d^2 V(x)}{dx^2}\right] = m(x)\, V(x)\frac{d^2 F(t)}{dt^2} \qquad (4.60)$$

Rearranging,

$$\frac{1}{m(x)\, V(x)}\frac{d^2}{dx^2}\left[EI(x)\frac{d^2 V(x)}{dx^2}\right] = -\frac{1}{F(t)}\frac{d^2 F(t)}{dt^2} \qquad (4.61)$$

For this equation to hold, both sides must be constant. Letting this constant be ω^2, the right-hand side may be written as

$$\frac{d^2F(t)}{dt^2} + \omega^2 F(t) = 0 \tag{4.62}$$

which has the same general solution as given by Equation (4.30). The left-hand side may be written as

$$\frac{d^2}{dx^2}\left[EI(x)\frac{d^2V(x)}{dx^2} \right] = \omega^2 m(x) \, V(x) \tag{4.63}$$

For the special case of a uniform beam (m is constant; EI is constant), Equation (4.63) becomes

$$EI\frac{d^4V(x)}{dx^4} - m\omega^2 V(x) = 0 \tag{4.64}$$

that is,

$$\frac{d^4V(x)}{dx^4} - \beta^4 \, V(x) = 0 \tag{4.65}$$

where

$$\beta^4 = \frac{m\omega^2}{EI} \tag{4.66}$$

The general solution of Equation (4.65) may be written in the form

$$V(x) = C_1 \sin \beta x + C_2 \cos \beta x + C_3 \sinh \beta x + C_4 \cosh \beta x \tag{4.67}$$

4.4.3.1 Simply supported beam

The boundary conditions (Equation (4.56)) may be written as

$$V(\xi) = C_1 \sin \beta\xi + C_2 \cos \beta\xi + C_3 \sinh \beta\xi + C_4 \cosh \beta\xi \tag{4.68a}$$

$$\left.\frac{d^2V(x)}{dx^2}\right|_{x=\xi} = \beta^2\left[-C_1 \sin\beta\xi - C_2 \cos\beta\xi + C_3 \sinh\beta\xi + C_4 \cosh\beta\xi \right] = 0 \tag{4.68b}$$

For $\xi = 0$, we have

$$V(0) = C_2 + C_4 = 0 \tag{4.69a}$$

$$\left. \frac{d^2V(x)}{dx^2} \right|_{\xi=0} = -C_2 + C_4 = 0 \tag{4.69b}$$

which gives the solutions

$$C_2 = C_4 = 0 \tag{4.70}$$

For $\xi = l$, and using the result (4.70), we have

$$V(l) = C_1 \sin\beta l + C_3 \sinh\beta l = 0 \tag{4.71a}$$

$$\left. \frac{d^2V(x)}{dx^2} \right|_{\xi=l} = -C_1 \sin\beta l + C_3 \sinh\beta l = 0 \tag{4.71b}$$

Adding the two expressions, we see that $C_3 \sinh\beta l$ is zero. Since βl is a positive quantity and $\sinh\beta l$ cannot be zero, it follows that

$$C_3 = 0 \tag{4.72a}$$

Equations (4.71a) and (4.71b) therefore reduce to the condition

$$C_1 \sin\beta l = 0 \tag{4.72b}$$

The constant C_1 cannot be zero (otherwise, there are no vibrations), so $\sin\beta l$ has to be zero. This gives the solutions

$$\beta_n l = n\pi \qquad (n = 1, 2, ..., \infty) \tag{4.73a}$$

that is,

$$\beta_n = \frac{n\pi}{l} \tag{4.73b}$$

From Equation (4.66), and using the result for β_n, we obtain

$$\omega_n = \beta_n^2 \sqrt{\frac{EI}{m}} = \frac{n^2\pi^2}{l^2}\sqrt{\frac{EI}{m}} = n^2\pi^2\sqrt{\frac{EI}{ml^4}} \tag{4.74}$$

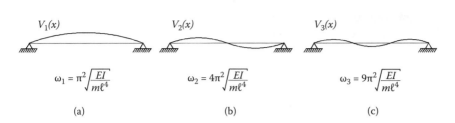

$$\omega_1 = \pi^2 \sqrt{\frac{EI}{m\ell^4}}$$

(a)

$$\omega_2 = 4\pi^2 \sqrt{\frac{EI}{m\ell^4}}$$

(b)

$$\omega_3 = 9\pi^2 \sqrt{\frac{EI}{m\ell^4}}$$

(c)

Figure 4.8 Mode shapes of the simply supported beam: (a) first mode, (b) second mode, (c) third mode.

Notice that the frequencies go up rapidly with increasing n. They are proportional to n^2, whereas in the case of the tensioned string and the rod experiencing axial vibrations, the frequencies have been seen to be proportional to just n. For the simply supported beam, the first three mode shapes, $V_1(x)$, $V_2(x)$ and $V_3(x)$, are depicted in Figure 4.8.

4.4.3.2 Cantilever beam

The boundary conditions (4.55) and (4.57) for a beam fixed at $x = 0$ and free at $x = l$, respectively, may be summarized as follows:

$$V(0) = 0 \tag{4.75a}$$

$$\left.\frac{dV(x)}{dx}\right|_{x=0} = 0 \tag{4.75b}$$

$$\left.\frac{d^2V(x)}{dx^2}\right|_{x=l} = 0 \tag{4.75c}$$

$$\left.\frac{d^3V(x)}{dx^3}\right|_{x=l} = 0 \tag{4.75d}$$

Imposing these conditions on the general solution (Equation (4.67)), we may readily establish the following relationship:

$$\cos\beta l \cosh\beta l = -1 \tag{4.76}$$

which, upon solving (numerically), gives the βl values for the various modes.

Expressing the other constants, C_2, C_3 and C_4, in terms of C_1 (by a process of successive elimination), and substituting the results into Equation (4.67), we finally obtain

$$V(x) = \left(\frac{C_1}{\sin \beta l - \sinh \beta l}\right)\left[(\sin \beta l - \sinh \beta l)(\sin \beta x - \sinh \beta x)\right. $$
$$\left. + (\cos \beta l + \cosh \beta l)(\cos \beta x - \cosh \beta x)\right] \tag{4.77}$$

which for the nth mode may be written in the form

$$V_n(x) = C_n\left[(\sin \beta_n l - \sinh \beta_n l)(\sin \beta_n x - \sinh \beta_n x)\right.$$
$$\left. + (\cos \beta_n l + \cosh \beta_n l)(\cos \beta_n x - \cosh \beta_n x)\right] \tag{4.78}$$

The amplitude C_n is indeterminate. However, it is not necessary to know it; it is the term in the square brackets that gives the mode shape.

As an exercise, the reader should derive in full the results given by Equations (4.76) and (4.77), proceed to evaluate numerically the first three circular natural frequencies ω_1, ω_2 and ω_3, and then plot the corresponding mode shapes $V_1(x)$, $V_2(x)$ and $V_3(x)$.

4.5 ORTHOGONALITY OF NATURAL MODES OF VIBRATION

We will prove the orthogonality of natural modes on the basis of the beam model, but clearly the result is general, and valid for all continuous systems.

For two distinct solutions $V_r(x)$ and $V_s(x)$ of the eigenvalue problem of the vibrating beam, we may write Equation (4.63) as follows:

$$\frac{d^2}{dx^2}\left[EI(x)\frac{d^2V_r(x)}{dx^2}\right] = \omega_r^2\, m(x)\, V_r(x) \tag{4.79a}$$

$$\frac{d^2}{dx^2}\left[EI(x)\frac{d^2V_s(x)}{dx^2}\right] = \omega_s^2\, m(x)\, V_s(x) \tag{4.79b}$$

Multiplying the first of these equations by $V_s(x)$, and integrating over the length of the beam, we obtain

$$\int_0^l V_s(x)\frac{d^2}{dx^2}\left[EI(x)\frac{d^2V_r(x)}{dx^2}\right]dx = \omega_r^2\int_0^l m(x)\,V_s(x)\,V_r(x)\,dx \tag{4.80}$$

The left-hand side may be successively integrated by parts as follows:

$$\int_0^l V_s(x)\frac{d^2}{dx^2}\left[EI(x)\frac{d^2V_r(x)}{dx^2}\right]dx \;\;\rightarrow\;\; \int_0^l V_s(x)\frac{d}{dx}\frac{d}{dx}\left[EI(x)\frac{d^2V_r(x)}{dx^2}\right]dx$$

$$\rightarrow V_s(x)\frac{d}{dx}\left[EI(x)\frac{d^2V_r(x)}{dx^2}\right] - \int_0^l \frac{d}{dx}\left[EI(x)\frac{d^2V_r(x)}{dx^2}\right]dV_s(x)$$

$$\rightarrow V_s(x)\frac{d}{dx}\left[EI(x)\frac{d^2V_r(x)}{dx^2}\right] - \int_0^l \frac{d}{dx}\left[EI(x)\frac{d^2V_r(x)}{dx^2}\right]\frac{dV_s(x)}{dx}dx$$

$$\rightarrow V_s(x)\frac{d}{dx}\left[EI(x)\frac{d^2V_r(x)}{dx^2}\right] - \int_0^l \frac{dV_s(x)}{dx}\frac{d}{dx}\left[EI(x)\frac{d^2V_r(x)}{dx^2}\right]dx$$

$$\rightarrow V_s(x)\frac{d}{dx}\left[EI(x)\frac{d^2V_r(x)}{dx^2}\right] - \left\{\frac{dV_s(x)}{dx}EI(x)\frac{d^2V_r(x)}{dx^2}\right.$$
$$\left. - \int_0^l EI(x)\frac{d^2V_r(x)}{dx^2}d\left(\frac{dV_s(x)}{dx}\right)\right\}$$

$$\rightarrow V_s(x)\frac{d}{dx}\left[EI(x)\frac{d^2V_r(x)}{dx^2}\right] - \left[\frac{dV_s(x)}{dx}EI(x)\frac{d^2V_r(x)}{dx^2}\right]$$
$$+ \int_0^l EI(x)\frac{d^2V_r(x)}{dx^2}\frac{d^2V_s(x)}{dx^2}dx$$

Using this result in Equation (4.80), we may finally write

$$\left\{V_s(x)\frac{d}{dx}\left[EI(x)\frac{d^2V_r(x)}{dx^2}\right]\right\}_0^l - \left[\frac{dV_s(x)}{dx}EI(x)\frac{d^2V_r(x)}{dx^2}\right]_0^l$$
$$+ \int_0^l EI(x)\frac{d^2V_r(x)}{dx^2}\frac{d^2V_s(x)}{dx^2}dx = \omega_r^2\int_0^l m(x)\,V_r(x)\,V_s(x)\,dx \tag{4.81}$$

Multiplying the second of Equations (4.79) by $V_r(x)$, and integrating over the length of the beam, we obtain

$$\int_0^l V_r(x)\frac{d^2}{dx^2}\left[EI(x)\frac{d^2V_s(x)}{dx^2}\right]dx = \omega_s^2\int_0^l m(x)\,V_r(x)\,V_s(x)\,dx \tag{4.82}$$

Integrating the left-hand side by parts as before, we finally obtain Equation (4.82) in the form:

$$
\left\{ V_r(x) \frac{d}{dx} \left[EI(x) \frac{d^2 V_s(x)}{dx^2} \right] \right\}_0^l - \left[\frac{dV_r(x)}{dx} EI(x) \frac{d^2 V_s(x)}{dx^2} \right]_0^l
$$

$$
+ \int_0^l EI(x) \frac{d^2 V_r(x)}{dx^2} \frac{d^2 V_s(x)}{dx^2} \, dx = \omega_s^2 \int_0^l m(x) \, V_r(x) \, V_s(x) \, dx
$$

(4.83)

Subtracting Equation (4.83) from Equation (4.81), we obtain

$$
\left(\omega_r^2 - \omega_s^2 \right) \int_0^l m(x) \, V_r(x) \, V_s(x) \, dx
$$

$$
= \left\{ V_s(x) \frac{d}{dx} \left[EI(x) \frac{d^2 V_r(x)}{dx^2} \right] \right\}_0^l - \left\{ V_r(x) \frac{d}{dx} \left[EI(x) \frac{d^2 V_s(x)}{dx^2} \right] \right\}_0^l
$$

$$
+ \left[\frac{dV_r(x)}{dx} EI(x) \frac{d^2 V_s(x)}{dx^2} \right]_0^l - \left[\frac{dV_s(x)}{dx} EI(x) \frac{d^2 V_r(x)}{dx^2} \right]_0^l
$$

(4.84)

Consider the right-hand side (RHS) of Equation (4.84). Either end of the beam is defined by the coordinate $x = \xi$, where ξ has the value 0 or l.

If an end $x = \xi$ is *fixed*, then

$$
V(\xi) = 0; \quad \left. \frac{dV(x)}{dx} \right|_{x=\xi} = 0
$$

(4.85a)

so that the RHS of Equation (4.84) reduces to zero *when evaluated for that end.*

If an end $x = \xi$ is *pinned*, then

$$
V(\xi) = 0; \quad \left. \frac{d^2 V(x)}{dx^2} \right|_{x=\xi} = 0
$$

(4.85b)

and the RHS of Equation (4.84) also reduces to zero when evaluated for that end.

If an end $x = \xi$ is *free*, then

$$
\left. \frac{d^2 V(x)}{dx^2} \right|_{x=\xi} = 0; \quad \left. \frac{d^3 V(x)}{dx^3} \right|_{x=\xi} = 0
$$

(4.85c)

so that the RHS of Equation (4.84) again reduces to zero when evaluated for that end.

For arbitrary support conditions at an end $x = \xi$ of the beam, it can be shown that the RHS of Equation (4.84) always reduces to zero when evaluated for that end. We conclude that the RHS of Equation (4.84), *when evaluated for the two ends* as defined by the limits $x = 0$ and $x = l$, always reduces to zero, irrespective of the combination of end conditions. Hence, Equation (4.84) reduces to

$$\left(\omega_r^2 - \omega_s^2\right) \int_0^l m(x)\, V_r(x)\, V_s(x)\, dx = 0 \tag{4.86}$$

For distinct values of ω_r and ω_s $(r \neq s)$, ω_r^2 is not equal to ω_s^2. It follows that

$$\int_0^l m(x)\, V_r(x)\, V_s(x)\, dx = 0 \tag{4.87}$$

showing that the eigenfunctions $V_r(x)$ and $V_s(x)$ are orthogonal with respect to the mass $m(x)$. This is an important result that will find application in evaluating the dynamic response of the beam to an applied excitation.

For $r = s$, the integral in Equation (4.86) is a positive quantity not equal to zero, and Equation (4.87) does not apply. As already stated, it is not possible to know the amplitude of the mode shapes, but we can *normalize* the mode shapes by setting this integral to unity, that is,

$$\int_0^l m(x)\, V_r(x)\, V_r(x)\, dx = 1 \tag{4.88}$$

To illustrate the concept of normalization of mode shapes, let us consider once again the example of the simply supported beam of uniform cross section. Since C_2, C_3 and C_4 were all found to be zero, $V_r(x)$ may be written down from Equation (4.67), with β_r being given by Equation (4.73b):

$$V_r(x) = C_{1r} \sin \beta_r x = C_{1r} \sin \frac{r\pi x}{l} \tag{4.89}$$

Normalizing the mode shapes, we can write

$$\int_0^l m\, V_r(x)\, V_r(x)\, dx = m C_{1r}^2 \int_0^l \sin^2\left(\frac{r\pi}{l}\right) x\, dx = 1 \tag{4.90}$$

Now, for any parameter η,

$$\int_0^l \sin^2 \eta x\, dx = \frac{1}{2}\left[x - \frac{\sin 2\eta x}{2\eta} \right]_0^l = \frac{1}{2}\left[l - \frac{\sin 2\eta l}{2\eta} \right] \tag{4.91}$$

In Equation (4.90), $\eta = r\pi/l$, so that

$$\sin 2\eta l = \sin\left(2\,\frac{r\pi}{l}\,l\right) = \sin 2\pi r = 0 \text{ for all } r \ (r = 1, 2, \ldots, \infty)$$

Thus, Equation (4.91) reduces to

$$\int_0^l \sin^2 \eta x \, dx = \frac{l}{2} \tag{4.92}$$

and Equation (4.90) in turn reduces to

$$m\, C_{1r}^2\, \frac{l}{2} = 1 \tag{4.93}$$

leading to the result

$$C_{1r} = \sqrt{\frac{2}{ml}} \tag{4.94}$$

Notice that the normalized amplitude C_{1r} (for the rth mode) is independent of the mode. It is the same for all modes ($r = 1, 2, \ldots, \infty$).

The normalized mode shapes may finally be written down, using result (4.94) in Equation (4.89), as follows:

$$V_r(x) = \sqrt{\frac{2}{ml}} \sin\frac{r\pi x}{l} \tag{4.95}$$

In Chapter 3, we saw that the eigenvectors for discrete systems are orthogonal not only with respect to the mass matrix, but also with respect to the stiffness matrix. For continuous systems, it seems reasonable to expect the eigenfunctions $V_r(x)$ and $V_s(x)$ to be also orthogonal with respect to the stiffness $EI(x)$. Let us investigate this. From the orthogonality of $V_r(x)$ and $V_s(x)$ with respect to the mass $m(x)$ (which was the result given by Equation (4.87)), we may write Equation (4.80) as follows:

$$\int_0^l V_s(x)\, \frac{d^2}{dx^2}\left[EI(x)\frac{d^2 V_r(x)}{dx^2}\right] dx = 0 \quad (r \neq s) \tag{4.96}$$

Earlier, we integrated by parts the left-hand side of the above equation. Using the result of that integration, we may recast Equation (4.96) in the form

$$V_s(x)\frac{d}{dx}\left[EI(x)\frac{d^2V_r(x)}{dx^2}\right]_0^l - \left[\frac{dV_s(x)}{dx}EI(x)\frac{d^2V_r(x)}{dx^2}\right]_0^l$$
$$+\int_0^l EI(x)\frac{d^2V_r(x)}{dx^2}\frac{d^2V_s(x)}{dx^2}dx = 0 \tag{4.97}$$

As argued earlier, the first two terms vanish for any conceivable combination of boundary conditions (fixed, pinned, free or arbitrary), so the above equation reduces to

$$\int_0^l EI(x)\frac{d^2V_r(x)}{dx^2}\frac{d^2V_s(x)}{dx^2}dx = 0 \quad (r \neq s) \tag{4.98}$$

showing that the *second derivatives* of the eigenfunctions (not the eigenfunctions themselves) are orthogonal with respect to the stiffness $EI(x)$.

For the normalized modes, we may write Equation (4.80) in the general form

$$\int_0^l V_s(x)\frac{d^2}{dx^2}\left[EI(x)\frac{d^2V_r(x)}{dx^2}\right]dx = \omega_r^2\delta_{rs} \tag{4.99}$$

where

$$\delta_{rs} = 1 \text{ for } r = s; \ \delta_{rs} = 0 \text{ for } r \neq s \tag{4.100}$$

4.6 DYNAMIC RESPONSE BY THE METHOD OF MODAL ANALYSIS

When the system is subject to a time-varying excitation, the ensuing dynamic response may be evaluated by the method of modal analysis encountered earlier in Chapter 3. In the present context of continuous systems, we will illustrate the method by reference to a beam, but of course, the procedure is equally applicable to other types of structures.

For simplicity, but without loss of generality, let us consider a beam of uniform properties, where EI is constant and m is also constant. We will assume that the beam is simply supported at both ends, and subjected to an external excitation $f(x,t)$ and the initial conditions

$$v(x,0) = v_0(x) \tag{4.101a}$$

$$\left.\frac{\partial v}{\partial t}\right|_{x,0} = \dot{v}_0(x) \tag{4.101b}$$

where $\dot{v}$ denotes the first derivative of v with respect to time, that is, the velocity. Equation (4.54) becomes

$$-EI \frac{\partial^4 v(x, t)}{\partial x^4} + f(x, t) = m \frac{\partial^2 v(x, t)}{\partial t^2} \tag{4.102}$$

The eigenvalue problem for $f(x,t) = 0$ has already been solved. We recap here the results for the natural circular frequency and normalized mode shape for the rth mode:

$$\omega_r = r^2 \pi^2 \sqrt{\frac{EI}{ml^4}} \tag{4.103a}$$

$$V_r(x) = \sqrt{\frac{2}{ml}} \sin \frac{r\pi x}{l} \tag{4.103b}$$

For the general problem represented by Equation (4.102), let us assume a solution in the form

$$v(x, t) = \sum_{r=1}^{\infty} V_r(x) \, q_r(t) \tag{4.104}$$

Substituting this into Equation (4.102), we obtain

$$\sum_{r=1}^{\infty} q_r(t) \, EI \frac{d^4 V_r(x)}{dx^4} + \sum_{r=1}^{\infty} \frac{d^2 q_r(t)}{dt^2} \, mV_r(x) = f(x, t) \tag{4.105}$$

Multiplying throughout by $V_s(x)$ and integrating over the beam, we get

$$\sum_{r=1}^{\infty} \left(\int_0^l V_s(x) \, EI \frac{d^4 V_r(x)}{dx^4} \, dx \right) q_r(t) + \sum_{r=1}^{\infty} \left(\int_0^l m \, V_r(x) \, V_s(x) \, dx \right) \frac{d^2 q_r(t)}{dt^2}$$
$$= \int_0^l f(x, t) \, V_s(x) \, dx \tag{4.106}$$

From the orthogonality relationships of Equations (4.87), (4.88) and (4.99), the first term of the above equation is nonzero only if $r = s$, and the second term is nonzero only if $r = s$. Substituting the values of the nonzero terms, Equation (4.106) becomes

$$\omega_r^2 \, q_r(t) + 1 \times \frac{d^2 q_r(t)}{dt^2} = \int_0^l f(x, t) \, V_r(x) \, dx \tag{4.107}$$

that is,

$$\frac{d^2 q_r(t)}{dt^2} + \omega_r^2 \, q_r(t) = F_r(t) \tag{4.108a}$$

or

$$\ddot{q}_r(t) + \omega_r^2 \, q_r(t) = F_r(t) \tag{4.108b}$$

where $F_r(t)$ are the generalized forces associated with the generalized coordinates $q_r(t)$:

$$F_r(t) = \int_0^l f(x, t) \, V_r(x) \, dx \tag{4.109}$$

Equation (4.108) has the same form as that for an undamped single degree-of-freedom system subjected to an exciting force (see Chapter 2). The general solution for $q_r(t)$ may therefore be written as (refer to Figure 4.9):

$$q_r(t) = \frac{1}{\omega_r} \int_0^t F_r(t') \sin \omega_r (t - t') \, dt' + q_{r0} \cos \omega_r t + \frac{\dot{q}_{r0}}{\omega_r} \sin \omega_r t \tag{4.110}$$

where $q_{r0} = q_r(0)$ and $\dot{q}_{r0} = \dot{q}_r(0)$ are the initial generalized coordinates and velocities. We need to express these in terms of actual initial displacements $v_0(x)$ and velocities $\dot{v}_0(x)$. From Equation (4.104), we can write

$$v(x, 0) = \sum_{r=1}^{\infty} V_r(x) \, q_r(0) = \sum_{r=1}^{\infty} V_r(x) \, q_{r0} \tag{4.111}$$

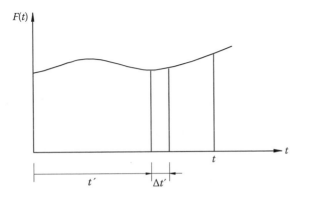

Figure 4.9 Variation of the generalized force F(t) over a time interval t.

Multiplying throughout by $mV_s(x)$ and integrating, we obtain

$$
\begin{aligned}
\int_0^l m\,V_s(x)\,v(x,0)\,dx &= \sum_{r=1}^\infty \left(\int_0^l m\,V_s(x)\,V_r(x)\,dx \right) q_{r0} \\
&= \left(\int_0^l m\,V_r(x)\,V_r(x)\,dx \right) q_{r0} = q_{r0}
\end{aligned}
\tag{4.112}
$$

(other terms vanishing due to orthogonality, and taking the normalized form of the mode shapes). Replacing $V_s(x)$ on the left-hand side by $V_r(x)$ (since $r = s$), and $v(x,0)$ by $v_0(x)$, we obtain the parameter q_{r0} as

$$
q_{r0} = \int_0^l m\,v_0(x)\,V_r(x)\,dx
\tag{4.113a}
$$

Similarly,

$$
\dot{q}_{r0} = \int_0^l m\,\dot{v}_0(x)\,V_r(x)\,dx
\tag{4.113b}
$$

The general response (Equation (4.104)) may finally be written down as follows:

$$
\begin{aligned}
v(x,t) = \sum_{r=1}^\infty V_r(x)\bigg[&\frac{1}{\omega_r}\int_0^t F_r(t')\sin\omega_r(t-t')\,dt' + q_{r0}\cos\omega_r t \\
&+ \frac{\dot{q}_{r0}}{\omega_r}\sin\omega_r t \bigg]
\end{aligned}
\tag{4.114}
$$

where $V_r(x)$ is given by Equation (4.103b), ω_r by Equation (4.103a), $F_r(t')$ by Equation (4.109) and $\{q_{r0}, \dot{q}_{r0}\}$ by Equations (4.113).

As a final example, let us consider the case of an applied loading in the form of a time-wise step function of magnitude p_o (constant over the full length of the beam) and duration t_o (Figure 4.10), given that the initial displacements $v_0(x)$ and initial velocities $\dot{v}_0(x)$ are zero. From Equations (4.113), it follows that q_{r0} and $\dot{q}_{r0}$ are also zero.

The loading is defined by the following functions (refer to Figure 4.10):

$$
f(x,t) = p_o \quad (0<t<t_o); \quad f(x,t) = 0 \quad (t>t_o)
\tag{4.115}
$$

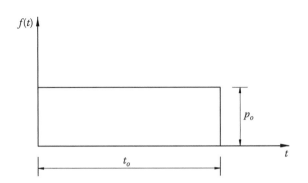

Figure 4.10 Applied loading in the form of a step function of magnitude p_o and duration t_o.

From Equation (4.109), and making use of relations (4.115) and (4.103b), we have

$$F_r(t) = \int_0^l p_o \sqrt{\frac{2}{ml}} \sin\left(\frac{r\pi x}{l}\right) dx = p_o \sqrt{\frac{2}{ml}} \frac{l}{r\pi} (1 - \cos r\pi) \qquad (4.116)$$

When r is even, the term in brackets vanishes; when r is odd, the term in brackets is equal to 2. Thus, we may write

$$F_r(t) = 2p_o \sqrt{\frac{2}{ml}} \frac{l}{r\pi} \qquad (r = 1, 3, 5, \ldots) \qquad (4.117)$$

With $F_r(t)$ now known, the general response follows from Equation (4.114), recalling that q_{r0} and $\dot{q}_{r0}$ are zero, and also making use of relations (4.103):

$$v(x,t) = \sum_{r=1}^{\infty} \sqrt{\frac{2}{ml}} \sin\frac{r\pi x}{l} \left[\frac{1}{\omega_r} 2p_o \sqrt{\frac{2}{ml}} \frac{l}{r\pi} \int_0^{t_o} \sin\omega_r(t - t') \, dt' \right]$$

$$= \frac{4p_o}{m\pi} \sum_{r=1}^{\infty} \frac{1}{r} \left(\sin\frac{r\pi x}{l} \right) \frac{1}{\omega_r^2} \left[\cos\omega_r(t - t') \right]_0^{t_o}$$

$$= \frac{4p_o}{m\pi} \sum_{r=1}^{\infty} \frac{1}{r} \frac{1}{r^4\pi^4} \frac{ml^4}{EI} \left(\sin\frac{r\pi x}{l} \right) \left[\cos\omega_r(t - t') \right]_0^{t_o}$$

$$= \frac{4p_o l^4}{\pi^5 EI} \sum_{r=1}^{\infty} \frac{1}{r^5} \left(\sin\frac{r\pi x}{l} \right) \left[\cos\omega_r(t - t_o) - \cos\omega_r t \right] (r = 1, 3, 5, \ldots) \ (4.118)$$

where ω_r is given by Equation (4.103a). Notice how the first mode dominates the motion (relative amplitudes of the modes are inversely proportional to r^5).

Chapter 5

Finite-element vibration analysis

5.1 THE FINITE-ELEMENT FORMULATION

5.1.1 Introduction

The finite-element method is a numerical formulation that divides the continuous structure into a number of discrete elements, where it is required that the displacements at the meeting points of the elements be compatible, and the internal forces at these points be in equilibrium. The procedure expresses the displacements at any arbitrary interior point of the element in terms of a finite number of displacements at the nodes of the element. Finite elements are typically one-, two- or three-dimensional, and examples are shown in Figure 5.1.

5.1.2 Displacements

Let the displacement field within the element be denoted by $\mathbf{u}$. For instance, in a plane–stress problem, where we might encounter rectangular elements of the type shown in Figure 5.1(b) or some other shapes of plane finite elements, we may write the generalized displacements as

$$\mathbf{u} = \left\{ \begin{array}{c} u(x, y) \\ v(x, y) \end{array} \right\} \tag{5.1}$$

where $u(x,y)$ and $v(x,y)$ denote displacements at the point defined by the coordinates $\{x,y\}$, in the x and the y directions, respectively.

Expressing such internal displacements in terms of the displacements at the nodes of the element, we may write

$$\mathbf{u} = \mathbf{Nd} \tag{5.2}$$

where $\mathbf{d}$ is a vector of the displacements at the nodes of the element, and $\mathbf{N}$ is a matrix of shape functions. In the case of the one-dimensional line

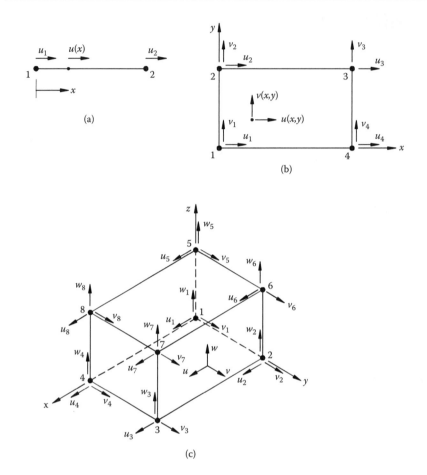

Figure 5.1 Various types of finite elements: (a) one-dimensional truss element, (b) two-dimensional plane–stress element, (c) three-dimensional solid brick element.

element of Figure 5.1(a), and the two-dimensional plane–stress element of Figure 5.1(b), we have

$$d = \left\{ \begin{array}{c} u_1 \\ u_2 \end{array} \right\}; \; d = \left\{ \begin{array}{cccccccc} u_1 & v_1 & u_2 & v_2 & u_3 & v_3 & u_4 & v_4 \end{array} \right\}^{\mathrm{T}} \qquad (5.3a, b)$$

5.1.3 Strains

The generalized strains ε at any point within the element may be written as

$$\varepsilon = L(u) \qquad (5.4)$$

where **L** is a linear operator. For instance, taking the example of the plane–stress element of Figure 5.1(b), we have

$$
\varepsilon = \left\{ \begin{array}{c} \varepsilon_x \\ \varepsilon_y \\ \gamma_{xy} \end{array} \right\} = \left\{ \begin{array}{c} \dfrac{\partial u}{\partial x} \\[2mm] \dfrac{\partial v}{\partial y} \\[2mm] \dfrac{\partial u}{\partial y} + \dfrac{\partial v}{\partial x} \end{array} \right\} = \left[\begin{array}{cc} \dfrac{\partial}{\partial x} & 0 \\[2mm] 0 & \dfrac{\partial}{\partial y} \\[2mm] \dfrac{\partial}{\partial y} & \dfrac{\partial}{\partial x} \end{array} \right] \left\{ \begin{array}{c} u \\ v \end{array} \right\} = L(\mathbf{u}) \qquad (5.5)
$$

where ε_x and ε_y are the direct strains in the x and y directions, respectively, and γ_{xy} is the shear strain with respect to the x and y directions; the operator L is the 3×2 matrix.

Substituting Equation (5.2) into Equation (5.4), we obtain

$$\varepsilon = L(\mathbf{Nd}) = L(\mathbf{N})\mathbf{d} = \mathbf{Bd} \qquad (5.6a)$$

where

$$\mathbf{B} = L(\mathbf{N}) \qquad (5.6b)$$

is the strain-displacement matrix.

5.1.4 Stresses

Generalized stresses **σ** may be related to generalized strains via the expression

$$\boldsymbol{\sigma} = \mathbf{D}\boldsymbol{\varepsilon} \qquad (5.7)$$

where **D** is a matrix of material stiffnesses. In the case of one-dimensional truss elements, where only axial resistance is relevant, **D** is simply the Young's modulus E of the material. For plane–stress problems, strains may be written in terms of stresses as follows:

$$\varepsilon_x = \frac{\sigma_x}{E} - v\frac{\sigma_y}{E}$$

$$\varepsilon_y = -v\frac{\sigma_x}{E} + \frac{\sigma_y}{E}$$

$$\gamma_{xy} = \frac{2(1+v)}{E}\tau_{xy}$$

that is,

$$
\left\{
\begin{array}{c}
\varepsilon_x \\
\varepsilon_y \\
\gamma_{xy}
\end{array}
\right\}
= \frac{1}{E}
\left[
\begin{array}{ccc}
1 & -\nu & 0 \\
-\nu & 1 & 0 \\
0 & 0 & 2(1+\nu)
\end{array}
\right]
\left\{
\begin{array}{c}
\sigma_x \\
\sigma_y \\
\tau_{xy}
\end{array}
\right\}
\tag{5.8a}
$$

or

$$
\varepsilon = \mathbf{C}\sigma
\tag{5.8b}
$$

Rearranging, we may write

$$
\sigma = \mathbf{C}^{-1}\varepsilon = \mathbf{D}\varepsilon
\tag{5.8c}
$$

Thus, for plane–stress problems,

$$
\mathbf{D} = \mathbf{C}^{-1} = \frac{E}{1-\nu^2}
\left[
\begin{array}{ccc}
1 & \nu & 0 \\
\nu & 1 & 0 \\
0 & 0 & \dfrac{1-\nu}{2}
\end{array}
\right]
\tag{5.9}
$$

5.1.5 Nodal actions and element stiffness matrix

We may think of nodal actions as actions that are statically equivalent to the system of loads (distributed and concentrated) acting on the element. Let us denote these by the vector **f**. Nodal actions may be actual forces or moments. They act in the same sense as the nodal displacements **d**. Let **g** denote the distributed loads per unit volume of material within the element, acting in the directions corresponding to the displacements **u**.

To obtain nodal actions **f** in terms of the loads acting on the element, we may use the concept of virtual work. Let us impose a virtual nodal displacement of magnitude δ**d** at the nodes. From Equations (5.2) and (5.6a), the resulting displacements and strains within the element are as follows:

$$
\delta \mathbf{u} = \mathbf{N}\delta \mathbf{d}
\tag{5.10a}
$$

$$
\delta \varepsilon = \mathbf{B}\delta \mathbf{d}
\tag{5.10b}
$$

The external work done by the nodal actions is the sum of the product of forces (or moments) and corresponding displacements (or rotations):

$$
\delta W_e = \delta \mathbf{d}^{\mathsf{T}} \mathbf{f}
\tag{5.11a}
$$

The internal work done *per unit volume* by the stresses and the distributed loads is given by

$$\delta W_i = \delta \varepsilon^\mathrm{T} \sigma - \delta u^\mathrm{T} g$$

$$= \delta d^\mathrm{T} B^\mathrm{T} \sigma - \delta d^\mathrm{T} N^\mathrm{T} g \text{ (making use of Equations (5.10b) and (5.10a))}$$

$$= \delta d^\mathrm{T}(B^\mathrm{T} \sigma - N^\mathrm{T} g) \tag{5.11b}$$

Equating the external work done to the total internal work done (the latter being obtained by integrating over the volume of the element), we obtain

$$f = \int_V B^\mathrm{T} \sigma \, dV - \int_V N^\mathrm{T} g \, dV \tag{5.12}$$

Making use of Equations (5.7) and (5.6) to eliminate first σ then ε, we obtain

$$f = \left(\int_V B^\mathrm{T} DB \, dV \right) d - \int_V N^\mathrm{T} g \, dV \tag{5.13}$$

that is,

$$f = Kd + b \tag{5.14}$$

where

$$K = \int_V B^\mathrm{T} DB \, dV \tag{5.15a}$$

and

$$b = - \int_V N^\mathrm{T} g \, dV \tag{5.15b}$$

The matrix K is the element stiffness matrix; b is the component of nodal actions contributed by the distributed loads over the volume of the element.

5.1.6 Dynamic formulation and element mass matrix

In a dynamic problem, the distributed body loads g are the inertial loads. From Equation (5.2), the accelerations of interior points may be expressed in terms of nodal accelerations as follows:

$$\ddot{u} = N\ddot{d} \tag{5.16}$$

The double-dot notation denotes second derivatives of the associated variables with respect to time. At any given interior point of the element, the inertial force per unit volume is given by the product of the mass per unit volume (ρ) and the acceleration $\ddot{u}$ at that point. Making use of expression (5.16) to eliminate $\ddot{u}$, the inertial force per unit volume is thus:

$$\rho\ddot{u} = \rho\mathbf{N}\ddot{\mathbf{d}} \tag{5.17}$$

As inertial force is opposite to the motion, $\mathbf{g}$ is therefore given by

$$\mathbf{g} = -\rho\mathbf{N}\ddot{\mathbf{d}} \tag{5.18}$$

The component of nodal actions due to distributed body loads (Equation (5.15b)) may thus be written in terms of inertial forces as follows:

$$\mathbf{b} = \left(\int_V \rho\mathbf{N}^T\mathbf{N}\,dV \right)\ddot{\mathbf{d}} \tag{5.19a}$$

that is,

$$\mathbf{b} = \mathbf{M}\ddot{\mathbf{d}} \tag{5.19b}$$

where

$$\mathbf{M} = \int_V \rho\mathbf{N}^T\mathbf{N}\,dV \tag{5.20}$$

The matrix $\mathbf{M}$ is the element mass matrix. In this case, it is called a *consistent mass matrix* because it is derived on the basis of the same shape functions as for the stiffness matrix $\mathbf{K}$.

Using Equation (5.19b) to eliminate $\mathbf{b}$ from Equation (5.14), we may write the equation of motion for the element as follows:

$$\mathbf{K}\mathbf{d} + \mathbf{M}\ddot{\mathbf{d}} = \mathbf{f} \tag{5.21}$$

For the assembly of all elements making up the structure, we may write the global equation of motion as

$$\mathbf{K}_g\mathbf{d}_g + \mathbf{M}_g\ddot{\mathbf{d}}_g = \mathbf{F}_g \tag{5.22a}$$

The excitations $\mathbf{F}_g$ are functions of time. Solving the above system of finite-element equations then yields the required dynamic response of the structure. In the case of free vibration, $\mathbf{F}_g$ is zero, and the above equation reduces to

$$\mathbf{K}_g\mathbf{d}_g + \mathbf{M}_g\ddot{\mathbf{d}}_g = 0 \tag{5.22b}$$

In the next sections, we derive the matrices $\mathbf{K}$ and $\mathbf{M}$ for a number of commonly used finite elements, and illustrate assembly and solution of the system equations of motion via a simple example.

5.2 STIFFNESS AND CONSISTENT MASS MATRICES FOR SOME COMMON FINITE ELEMENTS

5.2.1 Truss element

Figure 5.2 shows a two-noded truss element of length l, mass per unit length m and axial rigidity EA (where E is Young's modulus and A is the cross-sectional area). For this element, both m and EA are assumed to be constant. Let us denote the two ends of the element by 1 and 2, and the corresponding axial displacements by u_1 and u_2, respectively.

In the finite-element method, the usual approach is to assume a displacement field in the form of polynomial functions or some other suitable functions. For the present truss element, we may assume a linear displacement field in the form

$$u(x) = a_1 + a_2 x \tag{5.23}$$

This allows for both rigid-body motion and linear stretching. Now, at $x = 0$, we have $u = u_1$, and at $x = l$, we have $u = u_2$. Applying these two boundary conditions, we obtain

$$a_1 = u_1; \quad a_2 = \frac{u_2 - u_1}{l} \tag{5.24a, b}$$

Substituting these values of a_1 and a_2 into Equation (5.23), we may write

$$u(x) = u_1 + (u_2 - u_1)\frac{x}{l} = \left(1 - \frac{x}{l}\right)u_1 + \frac{x}{l}u_2$$

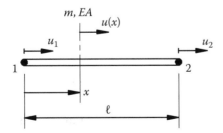

Figure 5.2 A two-noded truss element of length l.

that is,

$$u(x) = \left[\left(1 - \frac{x}{l}\right) \left(\frac{x}{l}\right) \right] \begin{Bmatrix} u_1 \\ u_2 \end{Bmatrix} = \mathbf{N}\mathbf{d} \tag{5.25}$$

where $\mathbf{N}$ is the matrix of shape functions, and $\mathbf{d}$ is the vector of nodal displacements:

$$\mathbf{N} = \left[\left(1 - \frac{x}{l}\right) \left(\frac{x}{l}\right) \right]; \quad \mathbf{d} = \begin{Bmatrix} u_1 \\ u_2 \end{Bmatrix} \tag{5.26}$$

Now for the one-dimensional deformations of a truss element, the strain ε_x, by definition, is given by

$$\varepsilon_x = \frac{\partial u}{\partial x} = \frac{\partial}{\partial x} u(x) = Lu(x), \text{ where } L = \frac{\partial}{\partial x} \tag{5.27}$$

Evaluating matrix $\mathbf{B}$ from Equation (5.6b), we obtain

$$\mathbf{B} = L(\mathbf{N}) = \frac{\partial}{\partial x} \left[\left(1 - \frac{x}{l}\right) \left(\frac{x}{l}\right) \right] = \left[-\frac{1}{l} \quad \frac{1}{l} \right] \tag{5.28}$$

so that

$$\mathbf{B}^\mathrm{T}\mathbf{D}\mathbf{B} = \begin{bmatrix} -\dfrac{1}{l} \\ \dfrac{1}{l} \end{bmatrix} E \left[-\frac{1}{l} \quad \frac{1}{l} \right] = \frac{E}{l^2} \begin{bmatrix} 1 & -1 \\ -1 & 1 \end{bmatrix} \tag{5.29}$$

The element stiffness matrix $\mathbf{K}$ follows as

$$\mathbf{K} = \int_V \mathbf{B}^\mathrm{T}\mathbf{D}\mathbf{B}\, dV = \int_0^l \mathbf{B}^\mathrm{T}\mathbf{D}\mathbf{B}\, A dx = \frac{E}{l^2} \begin{bmatrix} 1 & -1 \\ -1 & 1 \end{bmatrix} Al = \frac{EA}{l} \begin{bmatrix} 1 & -1 \\ -1 & 1 \end{bmatrix} \tag{5.30}$$

The consistent mass matrix $\mathbf{M}$ is evaluated from

$$\mathbf{M} = \int_V \rho \mathbf{N}^\mathrm{T}\mathbf{N}\, dV = \int_0^l \frac{m}{A} \mathbf{N}^\mathrm{T}\mathbf{N} A dx = m \int_0^l \mathbf{N}^\mathrm{T}\mathbf{N}\, dx \tag{5.31}$$

noting that the material density (ρ) is equal to the mass per unit length (m) divided by the cross-sectional area (A).

Now,

$$\mathbf{N}^T\mathbf{N} = \left[\begin{array}{c} \left(1-\dfrac{x}{l}\right) \\[2ex] \left(\dfrac{x}{l}\right) \end{array} \right] \left[\left(1-\dfrac{x}{l}\right)\left(\dfrac{x}{l}\right) \right] = \left[\begin{array}{cc} \left(1-\dfrac{2x}{l}+\dfrac{x^2}{l^2}\right) & \left(\dfrac{x}{l}-\dfrac{x^2}{l^2}\right) \\[3ex] \left(\dfrac{x}{l}-\dfrac{x^2}{l^2}\right) & \left(\dfrac{x^2}{l^2}\right) \end{array} \right] \qquad (5.32)$$

and

$$\int_0^l \mathbf{N}^T\mathbf{N}\,dx = \left[\begin{array}{cc} \left(x-\dfrac{x^2}{l}+\dfrac{x^3}{3l^2}\right) & \left(\dfrac{x^2}{2l}-\dfrac{x^3}{3l^2}\right) \\[3ex] \left(\dfrac{x^2}{2l}-\dfrac{x^3}{3l^2}\right) & \left(\dfrac{x^3}{3l^2}\right) \end{array} \right]_0^l = \left[\begin{array}{cc} \dfrac{l}{3} & \dfrac{l}{6} \\[2ex] \dfrac{l}{6} & \dfrac{l}{3} \end{array} \right] = \dfrac{l}{6}\left[\begin{array}{cc} 2 & 1 \\ 1 & 2 \end{array} \right]$$

$$(5.33)$$

Therefore,

$$\mathbf{M} = \frac{ml}{6}\left[\begin{array}{cc} 2 & 1 \\ 1 & 2 \end{array} \right] \qquad (5.34)$$

5.2.2 Beam element

Figure 5.3 shows a two-noded beam element of length l, mass per unit length m and flexural rigidity EI (where E is Young's modulus and I is the second moment of area). Both m and EI are assumed to be constant. The relevant generalized displacements in a beam are the transverse displacements, which we will denote by $v(x)$, and the rotations, which we will denote by $\theta(x)$. For ends 1 and 2 of the beam element, the corresponding

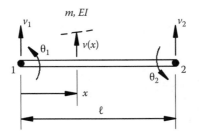

Figure 5.3 A two-noded beam element of length l.

sets of nodal displacements are therefore $\{v_1, \theta_1\}$ and $\{v_2, \theta_2\}$, respectively, as shown in the figure. For the transverse displacements, let us assume a cubic function of the form

$$v(x) = a_1 + a_2 x + a_3 x^2 + a_4 x^3 \tag{5.35}$$

The rotations follow as

$$\theta(x) = \frac{dv(x)}{dx} = a_2 + 2a_3 x + 3a_4 x^2 \tag{5.36}$$

To evaluate the constants a_1, a_2, a_3 and a_4, we impose the boundary conditions:

$$v(0) = v_1 \Rightarrow a_1 = v_1 \tag{5.37a}$$

$$\theta(0) = \theta_1 \Rightarrow a_2 = \theta_1 \tag{5.37b}$$

$$v(l) = v_2 \Rightarrow a_1 + a_2 l + a_3 l^2 + a_4 l^3 = v_2 \tag{5.37c}$$

$$\theta(l) = \theta_2 \Rightarrow a_2 + 2a_3 l + 3a_4 l^2 = \theta_2 \tag{5.37d}$$

leading to the solutions

$$a_1 = v_1 \tag{5.38a}$$

$$a_2 = \theta_1 \tag{5.38b}$$

$$a_3 = -\frac{3}{l^2} v_1 - \frac{2}{l} \theta_1 + \frac{3}{l^2} v_2 - \frac{1}{l} \theta_2 \tag{5.38c}$$

$$a_4 = \frac{2}{l^3} v_1 + \frac{1}{l^2} \theta_1 - \frac{2}{l^3} v_2 + \frac{1}{l^2} \theta_2 \tag{5.38d}$$

Substituting these values of a_1, a_2, a_3 and a_4 into Equation (5.35), we get

$$v(x) = v_1 + \theta_1 x + \left(-\frac{3}{l^2} v_1 - \frac{2}{l} \theta_1 + \frac{3}{l^2} v_2 - \frac{1}{l} \theta_2 \right) x^2 + \left(\frac{2}{l^3} v_1 + \frac{1}{l^2} \theta_1 - \frac{2}{l^3} v_2 + \frac{1}{l^2} \theta_2 \right) x^3$$

$$= \left(1 - 3\frac{x^2}{l^2} + 2\frac{x^3}{l^3} \right) v_1 + \left(x - 2\frac{x^2}{l} + \frac{x^3}{l^2} \right) \theta_1 + \left(3\frac{x^2}{l^2} - 2\frac{x^3}{l^3} \right) v_2$$

$$+ \left(-\frac{x^2}{l} + \frac{x^3}{l^2} \right) \theta_2$$

$$= N_1(x) v_1 + N_2(x) \theta_1 + N_3(x) v_2 + N_4(x) \theta_2 = \mathbf{Nd} \tag{5.39}$$

where

$$N_1(x) = 1 - 3\frac{x^2}{l^2} + 2\frac{x^3}{l^3} \tag{5.40a}$$

$$N_2(x) = x - 2\frac{x^2}{l} + \frac{x^3}{l^2} \tag{5.40b}$$

$$N_3(x) = 3\frac{x^2}{l^2} - 2\frac{x^3}{l^3} \tag{5.40c}$$

$$N_4(x) = -\frac{x^2}{l} + \frac{x^3}{l^2} \tag{5.40d}$$

and

$$\mathbf{N} = \begin{bmatrix} N_1(x) & N_2(x) & N_3(x) & N_4(x) \end{bmatrix} \tag{5.40e}$$

$$\mathbf{d} = \begin{Bmatrix} v_1 \\ \theta_1 \\ v_2 \\ \theta_2 \end{Bmatrix} \tag{5.40f}$$

The expressions $N_i(x)$ (for i = 1, 2, 3, 4) represent the shape functions of the beam element. Physically, $N_i(x)$ gives the displacements $v(x)$ over the beam when $\delta_i = 1$ and all other δs are zero, where $\delta_1 = v_1$, $\delta_2 = \theta_1$, $\delta_3 = v_2$ and $\delta_4 = \theta_2$. Figure 5.4 shows plots of the $N_i(x)$.

For a beam element, the strain-displacement matrix **B** now connects *curvatures* (rather than strains) with the displacements. For small displacements, the curvature κ is approximately equal to the second derivative of the displacement with respect to x. Thus,

$$\kappa(x) \approx \frac{d^2 v(x)}{dx^2} = \frac{d^2(\mathbf{Nd})}{dx^2} = L(\mathbf{Nd}) = L(\mathbf{N})\mathbf{d} = \mathbf{Bd} \tag{5.41}$$

where the operator L and the matrix **B** are given by

$$L = \frac{d^2}{dx^2}; \quad \mathbf{B} = L(\mathbf{N}) = \frac{d^2\mathbf{N}}{dx^2} \tag{5.42a, b}$$

Evaluating the elements of **B** by applying Equation (5.42b) on Equations (5.40), we obtain

$$B_1(x) = \frac{d^2 N_1(x)}{dx^2} = -\frac{6}{l^2} + \frac{12x}{l^3} \tag{5.43a}$$

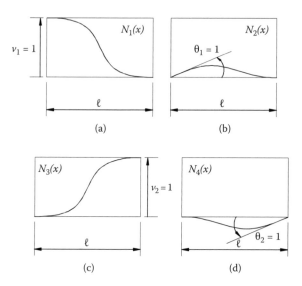

Figure 5.4 Shape functions for the two-noded beam element: (a) $N_1(x)$, (b) $N_2(x)$, (c) $N_3(x)$, (d) $N_4(x)$.

$$B_2(x) = \frac{d^2N_2(x)}{dx^2} = -\frac{4}{l} + \frac{6x}{l^2} \tag{5.43b}$$

$$B_3(x) = \frac{d^2N_3(x)}{dx^2} = \frac{6}{l^2} - \frac{12x}{l^3} \tag{5.43c}$$

$$B_4(x) = \frac{d^2N_4(x)}{dx^2} = -\frac{2}{l} + \frac{6x}{l^2} \tag{5.43d}$$

and so

$$\mathbf{B} = \begin{bmatrix} B_1(x) & B_2(x) & B_3(x) & B_4(x) \end{bmatrix} \tag{5.44}$$

The element stiffness matrix follows as

$$\mathbf{K} = \int_0^l \mathbf{B}^{\mathrm{T}}(EI)\mathbf{B}\, dx = EI \int_0^l \mathbf{B}^{\mathrm{T}}\mathbf{B}\, dx = EI \int_0^l \begin{bmatrix} B_1B_1 & B_1B_2 & B_1B_3 & B_1B_4 \\ B_2B_1 & B_2B_2 & B_2B_3 & B_2B_4 \\ B_3B_1 & B_3B_2 & B_3B_3 & B_3B_4 \\ B_4B_1 & B_4B_2 & B_4B_3 & B_4B_4 \end{bmatrix} dx \tag{5.45}$$

with B_1, B_2, B_3 and B_4 all being understood to be functions of x. The 16 coefficients of the stiffness matrix may be written as k_{ij} ($i = 1, 2, 3, 4; j = 1, 2, 3, 4$). Noting from Equation (5.45) that $k_{ij} = k_{ji}$, we only need to evaluate 10 distinct coefficients. As an example, k_{11} is evaluated as follows:

$$B_1 B_1 = \left(-\frac{6}{l^2}+\frac{12x}{l^3}\right)\left(-\frac{6}{l^2}+\frac{12x}{l^3}\right)=\frac{36}{l^4}-\frac{144x}{l^5}+\frac{144x^2}{l^6}$$

$$\int_0^l B_1 B_1\, dx = 36\frac{x}{l^4}-72\frac{x^2}{l^5}+48\frac{x^3}{l^6}\bigg|_0^l=\frac{12}{l^3}$$

$$k_{11}=EI\int_0^l B_1 B_1\, dx=12\left(\frac{EI}{l^3}\right)$$

Evaluating the rest of the k_{ij} in the same manner, we finally obtain

$$\mathbf{K}=\frac{EI}{l^3}\begin{bmatrix} 12 & 6l & -12 & 6l \\ 6l & 4l^2 & -6l & 2l^2 \\ -12 & -6l & 12 & -6l \\ 6l & 2l^2 & -6l & 4l^2 \end{bmatrix} \tag{5.46}$$

The consistent mass matrix for the beam element is evaluated on the basis of Equation (5.20), as follows:

$$\mathbf{M}=\int_V \rho\mathbf{N}^T\mathbf{N}\,dV=\int_0^l \rho\mathbf{N}^T\mathbf{N}\,A\,dx$$

$$=\rho A\int_0^l\begin{bmatrix} N_1N_1 & N_1N_2 & N_1N_3 & N_1N_4 \\ N_2N_1 & N_2N_2 & N_2N_3 & N_2N_4 \\ N_3N_1 & N_3N_2 & N_3N_3 & N_3N_4 \\ N_4N_1 & N_4N_2 & N_4N_3 & N_4N_4 \end{bmatrix}dx \tag{5.47}$$

As an example, the coefficient m_{11} of the matrix $\mathbf{M}$ is calculated as follows:

$$N_1N_1=\left(1-3\frac{x^2}{l^2}+2\frac{x^3}{l^3}\right)\left(1-3\frac{x^2}{l^2}+2\frac{x^3}{l^3}\right)=1-6\frac{x^2}{l^2}+4\frac{x^3}{l^3}+9\frac{x^4}{l^4}-12\frac{x^5}{l^5}+4\frac{x^6}{l^6}$$

$$\int_0^l N_1N_1\, dx = x-2\frac{x^3}{l^2}+\frac{x^4}{l^3}+\frac{9}{5}\frac{x^5}{l^4}-2\frac{x^6}{l^5}+\frac{4}{7}\frac{x^7}{l^6}\bigg|_0^l=\frac{156}{420}l$$

$$m_{11}=\rho A\int_0^l N_1N_1\, dx=156\frac{\rho Al}{420}$$

Evaluating the rest of the m_{ij} $(i = 1, 2, 3, 4; j = 1, 2, 3, 4)$ in the same way, we finally obtain:

$$M = \frac{\rho A l}{420} \begin{bmatrix} 156 & 22l & 54 & -13l \\ 22l & 4l^2 & 13l & -3l^2 \\ 54 & 13l & 156 & -22l \\ -13l & -3l^2 & -22l & 4l^2 \end{bmatrix} \tag{5.48}$$

5.2.3 Rectangular plane–stress element

Figure 5.5 shows a four-noded rectangular plane–stress element of dimensions a in the x direction and b in the y direction. At any interior point defined by the coordinates $\{x, y\}$, let the displacements in the x and y directions be denoted by $u(x, y)$ and $v(x, y)$, respectively. The nodal displacements are $\{u_i, v_i\}$, where $i = 1, 2, 3, 4$. For this element, we may assume a displacement field in the form

$$u(x, y) = a_1 + a_2 x + a_3 y + a_4 xy \tag{5.49a}$$

$$v(x, y) = a_5 + a_6 x + a_7 y + a_8 xy \tag{5.49b}$$

The constants $a_1, \ldots, a_8$ are obtained from the boundary conditions:

Point 1: $u(0, 0) = u_1$; $v(0, 0) = v_1$ (5.50a, b)

Point 2: $u(a, 0) = u_2$; $v(a, 0) = v_2$ (5.50c, d)

Point 3: $u(a, b) = u_3$; $v(a, b) = v_3$ (5.50e, f)

Point 4: $u(0, b) = u_4$; $v(0, b) = v_4$ (5.50g, h)

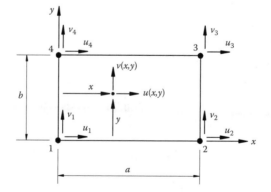

Figure 5.5 A four-noded rectangular plane–stress element of dimensions $a \times b$.

Substituting the solutions for $a_1, \ldots, a_8$ into Equations (5.49) and rearranging, we readily obtain expressions for interior displacements $u(x,y)$ and $v(x,y)$ in terms of the nodal displacements $\{u_1, u_2, u_3, u_4\}$ and $\{v_1, v_2, v_3, v_4\}$, which we may write in the form

$$u(x,y) = \begin{bmatrix} N_1(x,y) & N_2(x,y) & N_3(x,y) & N_4(x,y) \end{bmatrix} \begin{Bmatrix} u_1 \\ u_2 \\ u_3 \\ u_4 \end{Bmatrix} \tag{5.51a}$$

$$v(x,y) = \begin{bmatrix} N_1(x,y) & N_2(x,y) & N_3(x,y) & N_4(x,y) \end{bmatrix} \begin{Bmatrix} v_1 \\ v_2 \\ v_3 \\ v_4 \end{Bmatrix} \tag{5.51b}$$

where the shape functions $N_i(x,y)$ are the same for $u(x,y)$ and $v(x,y)$, and given by

$$N_1(x,y) = \left(1 - \frac{x}{a}\right)\left(1 - \frac{y}{b}\right) \tag{5.52a}$$

$$N_2(x,y) = \frac{x}{a}\left(1 - \frac{y}{b}\right) \tag{5.52b}$$

$$N_3(x,y) = \frac{xy}{ab} \tag{5.52c}$$

$$N_4(x,y) = \frac{y}{b}\left(1 - \frac{x}{a}\right) \tag{5.52d}$$

We may write the combined displacement field in the form

$$\begin{Bmatrix} u(x,y) \\ v(x,y) \end{Bmatrix} = \begin{bmatrix} N_1 & 0 & N_2 & 0 & N_3 & 0 & N_4 & 0 \\ 0 & N_1 & 0 & N_2 & 0 & N_3 & 0 & N_4 \end{bmatrix} \begin{Bmatrix} u_1 \\ v_1 \\ u_2 \\ v_2 \\ u_3 \\ v_3 \\ u_4 \\ v_4 \end{Bmatrix} = \mathbf{Nd} \tag{5.53}$$

where in the 2×8 matrix making up $\mathbf{N}$, the shape functions N_i ($i = 1, 2, 3, 4$) are functions of x and y as given in Equations (5.52). The four shape functions are plotted in Figure 5.6, where δ_i ($i = 1, 2, 3, 4$) equally represents u_i or v_i. We note that the distributions of the u and v displacements along any of the four edges of the element are linear. All sides remain straight when the element deforms, so compatibility of displacements between adjacent elements will always be fulfilled.

For plane–stress conditions, we use Equations (5.5) and (5.6b) to evaluate the strain-displacement matrix:

$$\mathbf{B} = L(\mathbf{N}) = \begin{bmatrix} \dfrac{\partial}{\partial x} & 0 \\[2mm] 0 & \dfrac{\partial}{\partial y} \\[2mm] \dfrac{\partial}{\partial y} & \dfrac{\partial}{\partial x} \end{bmatrix} \begin{bmatrix} N_1 & 0 & N_2 & 0 & N_3 & 0 & N_4 & 0 \\ 0 & N_1 & 0 & N_2 & 0 & N_3 & 0 & N_4 \end{bmatrix}$$

$$= \begin{bmatrix} \dfrac{\partial N_1}{\partial x} & 0 & \dfrac{\partial N_2}{\partial x} & 0 & \dfrac{\partial N_3}{\partial x} & 0 & \dfrac{\partial N_4}{\partial x} & 0 \\[2mm] 0 & \dfrac{\partial N_1}{\partial y} & 0 & \dfrac{\partial N_2}{\partial y} & 0 & \dfrac{\partial N_3}{\partial y} & 0 & \dfrac{\partial N_4}{\partial y} \\[2mm] \dfrac{\partial N_1}{\partial y} & \dfrac{\partial N_1}{\partial x} & \dfrac{\partial N_2}{\partial y} & \dfrac{\partial N_2}{\partial x} & \dfrac{\partial N_3}{\partial y} & \dfrac{\partial N_3}{\partial x} & \dfrac{\partial N_4}{\partial y} & \dfrac{\partial N_4}{\partial x} \end{bmatrix} \quad (5.54)$$

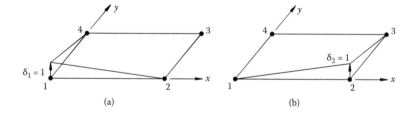

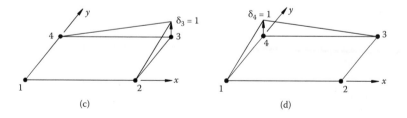

Figure 5.6 Shape functions for the four-noded rectangular plane–stress element: (a) $N_1(x,y)$, (b) $N_2(x,y)$, (c) $N_3(x,y)$, (d) $N_4(x,y)$.

Evaluating the partial derivatives in the above expression, we obtain

$$
\mathbf{B} = \begin{bmatrix}
-\dfrac{1}{a}\left(1-\dfrac{y}{b}\right) & 0 & \dfrac{1}{a}\left(1-\dfrac{y}{b}\right) & 0 & \dfrac{y}{ab} & 0 & -\dfrac{y}{ab} & 0 \\[2ex]
0 & -\dfrac{1}{b}\left(1-\dfrac{x}{a}\right) & 0 & -\dfrac{x}{ab} & 0 & \dfrac{x}{ab} & 0 & \dfrac{1}{b}\left(1-\dfrac{x}{a}\right) \\[2ex]
-\dfrac{1}{b}\left(1-\dfrac{x}{a}\right) & -\dfrac{1}{a}\left(1-\dfrac{y}{b}\right) & -\dfrac{x}{ab} & \dfrac{1}{a}\left(1-\dfrac{y}{b}\right) & \dfrac{x}{ab} & \dfrac{y}{ab} & \dfrac{1}{b}\left(1-\dfrac{x}{a}\right) & -\dfrac{y}{ab}
\end{bmatrix}
$$

$$(5.55)$$

Assuming the thickness t is constant over the surface of the element, $\mathbf{K}$ may be evaluated from Equation (5.15a), with $\mathbf{D}$ (for plane stress) being given by Equation (5.9):

$$
\mathbf{K} = \int_V \mathbf{B}^\mathrm{T}\mathbf{D}\mathbf{B}\, dV = \int_A \mathbf{B}^\mathrm{T}\mathbf{D}\mathbf{B}\, t dA = t\int_0^b \int_0^a \mathbf{B}^\mathrm{T}\mathbf{D}\mathbf{B}\, dx\, dy
$$

$$
= \frac{Et}{1-v^2}\int_0^b \int_0^a \mathbf{B}^\mathrm{T}
\begin{bmatrix}
1 & v & 0 \\
v & 1 & 0 \\
0 & 0 & \dfrac{1-v}{2}
\end{bmatrix}
\mathbf{B}\, dx\, dy
$$

$$(5.56)$$

where $\mathbf{B}$ is given by Equation (5.55) above.

The element stiffness matrix $\mathbf{K}$ is clearly 8×8. We may readily evaluate its coefficients by first performing the matrix multiplications in the integral, and then integrating each term of the ensuing 8×8 matrix between the given limits. For instance, denoting by β_{11} the element in the first row and the first column of the 8×8 matrix in the integral, we use the first column of $\mathbf{B}$ to evaluate β_{11}, as follows:

$$
\beta_{11} = \begin{bmatrix} -\dfrac{1}{a}\left(1-\dfrac{y}{b}\right) & 0 & -\dfrac{1}{b}\left(1-\dfrac{x}{a}\right) \end{bmatrix}
\begin{bmatrix}
1 & v & 0 \\
v & 1 & 0 \\
0 & 0 & \dfrac{1-v}{2}
\end{bmatrix}
\begin{bmatrix}
-\dfrac{1}{a}\left(1-\dfrac{y}{b}\right) \\[1.5ex]
0 \\[1.5ex]
-\dfrac{1}{b}\left(1-\dfrac{x}{a}\right)
\end{bmatrix}
$$

$$
= \begin{bmatrix} -\dfrac{1}{a}\left(1-\dfrac{y}{b}\right) & -\dfrac{v}{a}\left(1-\dfrac{y}{b}\right) & -\dfrac{1-v}{2b}\left(1-\dfrac{x}{a}\right) \end{bmatrix}
\begin{bmatrix}
-\dfrac{1}{a}\left(1-\dfrac{y}{b}\right) \\[1.5ex]
0 \\[1.5ex]
-\dfrac{1}{b}\left(1-\dfrac{x}{a}\right)
\end{bmatrix}
$$

$$= \frac{1}{a^2}\left(1-\frac{y}{b}\right)\left(1-\frac{y}{b}\right) + \frac{1-v}{2b^2}\left(1-\frac{x}{a}\right)\left(1-\frac{x}{a}\right) \tag{5.57}$$

Integrating β_{11}, we obtain

$$\int_0^b\int_0^a \beta_{11}\,dx\,dy = \int_0^b\int_0^a \frac{1}{a^2}\left(1-\frac{2y}{b}+\frac{y^2}{b^2}\right) + \frac{1-v}{2b^2}\left(1-\frac{2x}{a}+\frac{x^2}{a^2}\right)dx\,dy$$

$$= \left[\left[\frac{x}{a^2}\left(y-\frac{y^2}{b}+\frac{y^3}{3b^2}\right)+\left(\frac{1-v}{2b^2}\right)y\left(x-\frac{x^2}{a}+\frac{x^3}{3a^2}\right)\right]_0^a\right]_0^b$$

$$= \frac{b}{3a}+\left(\frac{1-v}{6}\right)\frac{a}{b} \tag{5.58}$$

Therefore, the element k_{11} of the stiffness matrix is, from Equations (5.56) and (5.58),

$$k_{11} = \frac{Et}{1-v^2}\int_0^b\int_0^a \beta_{11}\,dx\,dy = \frac{Et}{1-v^2}\left[\frac{b}{3a}+\left(\frac{1-v}{6}\right)\frac{a}{b}\right] \tag{5.59}$$

The rest of the coefficients of $\mathbf{K}$ are evaluated in a similar manner, taking advantage of the property $k_{ij} = k_{ji}$. The full stiffness matrix (symmetric about the diagonal) is obtained as

$$\mathbf{K} = \frac{Et}{24(1-v^2)}\begin{bmatrix} k_1 & \cdot & \cdot & \cdot & \cdot & \cdot & \cdot & \cdot \\ k_2 & k_3 & \cdot & \cdot & \cdot & \cdot & \cdot & \cdot \\ k_4 & -k_5 & k_1 & \cdot & \cdot & \cdot & \cdot & \cdot \\ k_5 & k_6 & -k_2 & k_3 & \cdot & \cdot & \cdot & \cdot \\ -0.5k_1 & -k_2 & k_7 & -k_5 & k_1 & \cdot & \cdot & \cdot \\ -k_2 & -0.5k_3 & k_5 & k_8 & k_2 & k_3 & \cdot & \cdot \\ k_7 & k_5 & -0.5k_1 & k_2 & k_4 & -k_5 & k_1 & \cdot \\ -k_5 & k_8 & k_2 & -0.5k_3 & k_5 & k_6 & -k_2 & k_3 \end{bmatrix} \tag{5.60}$$

where the nondimensional parameters $k_1, \ldots, k_8$ are defined as follows:

$$k_1 = 8\frac{b}{a}+4(1-v)\frac{a}{b} \tag{5.61a}$$

$$k_2 = 6v + 3(1-v) = 3(1+v) \tag{5.61b}$$

$$k_3 = 8\frac{a}{b}+4(1-v)\frac{b}{a} \tag{5.61c}$$

$$k_4 = -8\frac{b}{a} + 2(1-\nu)\frac{a}{b} \qquad (5.61d)$$

$$k_5 = 6\nu - 3(1-\nu) = -3(1-3\nu) \qquad (5.61e)$$

$$k_6 = 4\frac{a}{b} - 4(1-\nu)\frac{b}{a} \qquad (5.61f)$$

$$k_7 = 4\frac{b}{a} - 4(1-\nu)\frac{a}{b} \qquad (5.61g)$$

$$k_8 = -8\frac{a}{b} + 2(1-\nu)\frac{b}{a} \qquad (5.61h)$$

On the basis of Equation (5.20), we evaluate the consistent mass matrix **M** as follows, with **N** being the 2×8 matrix of Equation (5.53):

$$\mathbf{M} = \rho t \int_0^b \int_0^a \mathbf{N}^T \mathbf{N}\, dx\, dy$$

$$= \rho t \int_0^b \int_0^a \begin{bmatrix} N_1 N_1 & & N_1 N_2 & & N_1 N_3 & & N_1 N_4 & \\ & N_1 N_1 & & N_1 N_2 & & N_1 N_3 & & N_1 N_4 \\ N_2 N_1 & & N_2 N_2 & & N_2 N_3 & & N_2 N_4 & \\ & N_2 N_1 & & N_2 N_2 & & N_2 N_3 & & N_2 N_4 \\ N_3 N_1 & & N_3 N_2 & & N_3 N_3 & & N_3 N_4 & \\ & N_3 N_1 & & N_3 N_2 & & N_3 N_3 & & N_3 N_4 \\ N_4 N_1 & & N_4 N_2 & & N_4 N_3 & & N_4 N_4 & \\ & N_4 N_1 & & N_4 N_2 & & N_4 N_3 & & N_4 N_4 \end{bmatrix} dx\, dy$$

$$(5.62)$$

For example, to evaluate m_{11}, we first evaluate $N_1 N_1$ and its integral:

$$N_1 N_1 = \left(1 - \frac{x}{a}\right)\left(1 - \frac{x}{a}\right)\left(1 - \frac{y}{b}\right)\left(1 - \frac{y}{b}\right) = \left(1 - \frac{2x}{a} + \frac{x^2}{a^2}\right)\left(1 - \frac{2y}{b} + \frac{y^2}{b^2}\right) \qquad (5.63)$$

$$\int_0^b \int_0^a N_1 N_1\, dx\, dy = \left(x - \frac{x^2}{a} + \frac{x^3}{3a^2}\right)\left(y - \frac{y^2}{b} + \frac{y^3}{3b^2}\right)\Bigg|_0^a \Bigg|_0^b = \frac{a}{3}\frac{b}{3} = \frac{ab}{9} \qquad (5.64)$$

The coefficient m_{11} immediately follows:

$$m_{11} = \rho t \int_0^b \int_0^a N_1 N_1\, dx\, dy = \frac{\rho t a b}{9} \qquad (5.65)$$

Performing the evaluations for all the coefficients of the consistent mass matrix $\mathbf{M}$ of the element, we obtain the final result:

$$\mathbf{M} = \frac{\rho tab}{36} \begin{bmatrix} 4 & 0 & 2 & 0 & 1 & 0 & 2 & 0 \\ 0 & 4 & 0 & 2 & 0 & 1 & 0 & 2 \\ 2 & 0 & 4 & 0 & 2 & 0 & 1 & 0 \\ 0 & 2 & 0 & 4 & 0 & 2 & 0 & 1 \\ 1 & 0 & 2 & 0 & 4 & 0 & 2 & 0 \\ 0 & 1 & 0 & 2 & 0 & 4 & 0 & 2 \\ 2 & 0 & 1 & 0 & 2 & 0 & 4 & 0 \\ 0 & 2 & 0 & 1 & 0 & 2 & 0 & 4 \end{bmatrix} \qquad (5.66)$$

5.3 ASSEMBLY OF THE SYSTEM EQUATIONS OF MOTION

5.3.1 Procedure in brief

A detailed description of the process of assembly of the element matrices into the global (or system) matrices may be seen in various textbooks devoted to the finite-element method [1–4].

To begin with, a common coordinate system for the full structure needs to be chosen, and by applying appropriate transformation matrices, the individual element matrices (in terms of local coordinates) are expressed in terms of the global coordinates.

The nodes of the structure are typically the points at which the various elements are connected together. We may think of the stiffness of a structural node as being equal to the sum of the corresponding stiffnesses of the elements meeting at that node, and then assemble the structure stiffness matrix by simply superimposing the element matrices in accordance with the way the elements are joined up. Typically, we end up with a much bigger matrix representing the stiffness of the structure as a whole, and reflecting all the freedoms (in global coordinates) of the structure at its various nodes. If some of the freedoms are constrained as a result of the way the structure is supported, then we have to modify the assembled stiffness matrix accordingly.

The global mass matrix for the structure is assembled in a similar fashion.

5.3.2 Illustration of assembly

Let us illustrate the procedure outlined above by reference to the beam shown in Figure 5.7. The beam is divided into two two-noded beam finite elements, which we will denote as elements 1 and 2, the corresponding lengths being l_1 and l_2, respectively.

The system has three structural nodes, which are labelled 1, 2 and 3, as shown. Node 1 is the left-hand end of the beam, node 2 is the connection

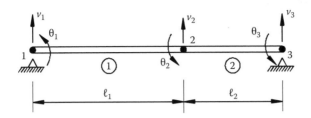

Figure 5.7 Illustrative example: A beam divided into two finite elements.

point of the two elements, and node 3 is the right-hand end of the beam. Each of these structural nodes potentially has two degrees of freedom (a vertical translation and a rotation), giving six degrees of freedom for the structure as a whole, as shown in the figure.

Using subscripts 1 and 2 to denote variables of elements 1 and 2, we may define two stiffness parameters $\{s_1, s_2\}$ and two mass parameters $\{m_1, m_2\}$ for these elements, as follows:

$$s_1 = \frac{EI_1}{l_1^3}; s_2 = \frac{EI_2}{l_2^3}; m_1 = \frac{\rho A_1 l_1}{420}; m_2 = \frac{\rho A_2 l_2}{420} \tag{5.67}$$

These parameters are the multipliers in front of the element stiffness matrix (Equation (5.46)) and the element mass matrix (Equation (5.48)).

Since the structure has six degrees of freedom, its assembled global matrices will be 6 × 6 matrices. The components of the global stiffness matrix corresponding to the global freedoms v_2 and θ_2 (at node 2) are the sum of the contributions coming from the right end of element 1 and the left end of element 2.

Using the parameters s_1 and s_2 for elements 1 and 2, respectively, we may therefore assemble the global stiffness matrix $\mathbf{K}_g$ as follows:

$$\mathbf{K}_g \mathbf{d}_g = \begin{bmatrix} 12s_1 & 6l_1s_1 & -12s_1 & 6l_1s_1 & & \\ 6l_1s_1 & 4l_1^2 s_1 & -6l_1s_1 & 2l_1^2 s_1 & & \\ -12s_1 & -6l_1s_1 & \begin{matrix}12s_1\\+12s_2\end{matrix} & \begin{matrix}-6l_1s_1\\+6l_2s_2\end{matrix} & -12s_2 & 6l_2s_2 \\ 6l_1s_1 & 2l_1^2 s_1 & \begin{matrix}-6l_1s_1\\+6l_2s_2\end{matrix} & \begin{matrix}4l_1^2 s_1\\+4l_2^2 s_2\end{matrix} & -6l_2s_2 & 2l_2^2 s_2 \\ & & -12s_2 & -6l_2s_2 & 12s_2 & -6l_2s_2 \\ & & 6l_2s_2 & 2l_2^2 s_2 & -6l_2s_2 & 4l_2^2 s_2 \end{bmatrix} \begin{Bmatrix} v_1 \\ \theta_1 \\ v_2 \\ \theta_2 \\ v_3 \\ \theta_3 \end{Bmatrix}$$

$$\tag{5.68}$$

Similarly, the global mass matrix $\mathbf{M}_g$ is assembled as follows:

$$\mathbf{M}_g\ddot{\mathbf{d}}_g = \begin{bmatrix} 156m_1 & 22l_1m_1 & 54m_1 & -13l_1m_1 & & \\ 22l_1m_1 & 4l_1^2m_1 & 13l_1m_1 & -3l_1^2m_1 & & \\ 54m_1 & 13l_1m_1 & \begin{array}{c}156m_1\\+156m_2\end{array} & \begin{array}{c}-22l_1m_1\\+22l_2m_2\end{array} & 54m_2 & -13l_2m_2 \\ -13l_1m_1 & -3l_1^2m_1 & \begin{array}{c}-22l_1m_1\\+22l_2m_2\end{array} & \begin{array}{c}4l_1^2m_1\\+4l_2^2m_2\end{array} & 13l_2m_2 & -3l_2^2m_2 \\ & & 54m_2 & 13l_2m_2 & 156m_2 & -22l_2m_2 \\ & & -13l_2m_2 & -3l_2^2m_2 & -22l_2m_2 & 4l_2^2m_2 \end{bmatrix} \begin{Bmatrix} \ddot{v}_1 \\ \ddot{\theta}_1 \\ \ddot{v}_2 \\ \ddot{\theta}_2 \\ \ddot{v}_3 \\ \ddot{\theta}_3 \end{Bmatrix}$$

$$(5.69)$$

The above matrices $\mathbf{K}_g$ and $\mathbf{M}_g$ assume all six freedoms $\{v_1,\theta_1\}$, $\{v_2,\theta_2\}$ and $\{v_3,\theta_3\}$ are free. They do not take into account the support conditions of the beam.

5.3.3 Accounting for support conditions

Let us consider the cases when the ends of the beam (nodes 1 and 3) are (1) simply supported and (2) clamped.

1. If $v_1 = 0$ and $v_3 = 0$ (beam simply supported at both ends), we suppress rows 1 and 5, as well as columns 1 and 5, so that the eigenvalue equation of the system becomes

$$\left[\begin{bmatrix} 4l_1^2s_1 & -6l_1s_1 & 2l_1^2s_1 & 0 \\ -6l_1s_1 & \begin{array}{c}12s_1\\+12s_2\end{array} & \begin{array}{c}-6l_1s_1\\+6l_2s_2\end{array} & 6l_2s_2 \\ 2l_1^2s_1 & \begin{array}{c}-6l_1s_1\\+6l_2s_2\end{array} & \begin{array}{c}4l_1^2s_1\\+4l_2^2s_2\end{array} & 2l_2^2s_2 \\ 0 & 6l_2s_2 & 2l_2^2s_2 & 4l_2^2s_2 \end{bmatrix} \right.$$

$$(5.70)$$

$$\left. -\omega^2 \begin{bmatrix} 4l_1^2m_1 & 13l_1m_1 & -3l_1^2m_1 & 0 \\ 13l_1m_1 & \begin{array}{c}156m_1\\+156m_2\end{array} & \begin{array}{c}-22l_1m_1\\+22l_2m_2\end{array} & -13l_2m_2 \\ -3l_1^2m_1 & \begin{array}{c}-22l_1m_1\\+22l_2m_2\end{array} & \begin{array}{c}4l_1^2m_1\\+4l_2^2m_2\end{array} & -3l_2^2m_2 \\ 0 & -13l_2m_2 & -3l_2^2m_2 & 4l_2^2m_2 \end{bmatrix} \right] \begin{Bmatrix} \theta_1 \\ v_2 \\ \theta_2 \\ \theta_3 \end{Bmatrix} = 0$$

or

$$\left[\mathbf{K}_g - \lambda \mathbf{M}_g \right] \mathbf{d}_g = 0 \tag{5.71}$$

where

$$\lambda = \omega^2 \tag{5.72}$$

Setting the determinant $\left| \mathbf{K}_g - \lambda \mathbf{M}_g \right|$ to zero results in a fourth-degree characteristic equation in λ, which upon solving yields four eigenvalues $\{\lambda_1, \lambda_2, \lambda_3, \lambda_4\}$, and hence the natural circular frequencies $\{\omega_1, \omega_2, \omega_3, \omega_4\}$ of the system.

2. If $\{v_1 = 0; \theta_1 = 0\}$ and $\{v_3 = 0; \theta_3 = 0\}$ (beam fixed at both ends), we suppress rows $\{1, 2, 5, 6\}$ and columns $\{1, 2, 5, 6\}$ corresponding to the freedoms $\{v_1, \theta_1, v_3, \theta_3\}$, so that the eigenvalue equation of the system becomes

$$\left[\begin{bmatrix} 12s_1 + 12s_2 & -6l_1s_1 + 6l_2s_2 \\ \hline -6l_1s_1 + 6l_2s_2 & 4l_1^2 s_1 + 4l_2^2 s_2 \end{bmatrix} - \omega^2 \begin{bmatrix} 156m_1 + 156m_2 & -22l_1m_1 + 22l_2m_2 \\ \hline -22l_1m_1 + 22l_2m_2 & 4l_1^2 m_1 + 4l_2^2 m_2 \end{bmatrix} \right] \begin{Bmatrix} v_2 \\ \theta_2 \end{Bmatrix} = 0 \tag{5.73}$$

Setting the determinant $\left| \mathbf{K}_g - \lambda \mathbf{M}_g \right|$ to zero results in a second-degree characteristic equation in λ, which upon solving yields two eigenvalues $\{\lambda_1, \lambda_2\}$, and hence the natural circular frequencies $\{\omega_1, \omega_2\}$ of the system. In this and the previous example, notice that there are as many natural frequencies of the system as there are unrestrained degrees of freedom, a feature of all dynamic finite-element models.

5.3.4 Numerical example

For the example of the beam fixed at both ends, let

$$l_1 = l_2 = l; \; A_1 = A_2 = A; \; I_1 = I_2 = I \tag{5.74}$$

so that

$$k_1 = k_2 = \frac{EI}{l^3} = k; \; m_1 = m_2 = \frac{\rho A l}{420} = m_o \tag{5.75}$$

The determinant of Equation (5.73), when equated to zero, becomes

$$\left| \begin{bmatrix} 24k & 0 \\ 0 & 8kl^2 \end{bmatrix} - \begin{bmatrix} 312m_o\lambda & 0 \\ 0 & 8m_o l^2 \lambda \end{bmatrix} \right| = 0 \tag{5.76a}$$

that is,

$$\begin{vmatrix} 24k - 312m_o\lambda & 0 \\ 0 & 8kl^2 - 8m_ol^2\lambda \end{vmatrix} = 0 \qquad (5.76b)$$

Expanding this, we get

$$(24k - 312m_o\lambda)(8kl^2 - 8m_ol^2\lambda) = 0 \qquad (5.77a)$$

that is,

$$\left(2496m_o^2l^2\right)\lambda^2 - \left(2688m_okl^2\right)\lambda + 192k^2l^2 = 0 \qquad (5.77b)$$

leading to the solutions

$$\lambda = \frac{k}{m_o} \text{ and } \lambda = \frac{1}{13}\frac{k}{m_o} \qquad (5.78a)$$

which we may write (smaller one first) as

$$\lambda_1 = 0.07692\frac{k}{m_o}; \; \lambda_2 = \frac{k}{m_o} \qquad (5.78b)$$

The corresponding values of the natural circular frequencies follow from Equation (5.72):

$$\omega_1 = 0.27735\sqrt{\frac{k}{m_o}}; \; \omega_2 = \sqrt{\frac{k}{m_o}} \qquad (5.79)$$

From Equations (5.75), and replacing ρA by m (mass per unit length), we have

$$\frac{k}{m_o} = \frac{EI}{l^3}\frac{420}{\rho Al} = \frac{420EI}{ml^4} \qquad (5.80)$$

Using this result in Equations (5.79), we finally obtain

$$\omega_1 = 5.684\sqrt{\frac{EI}{ml^4}}; \quad \omega_2 = 20.494\sqrt{\frac{EI}{ml^4}}$$

which are the natural circular frequencies of vibration of the beam fixed at both ends.

It should be noted that the finite-element method is an approximate method. The larger the number of elements we take (that is, the finer the finite-element mesh), the closer will the obtained results approach the exact solution (as obtained by the methods of Chapter 4).

REFERENCES

1. O.C. Zienkiewicz and R.L. Taylor. *The Finite Element Method: Basic Formulation and Linear Problems*. Vol. 1. McGraw-Hill, London, 1989.
2. R.D. Cook. *Concepts and Applications of Finite Element Analysis*. John Wiley, New York, 1981.
3. K.J. Bathe. *Finite Element Procedures in Engineering Analysis*. Prentice-Hall, Englewood Cliffs, NJ, 1982.
4. T.Y. Yang. *Finite Element Structural Analysis*. Prentice-Hall, Englewood Cliffs, NJ, 1986.

Part II

Group-theoretic formulations

Chapter 6

Basic concepts of symmetry groups and representation theory

6.1 SYMMETRY GROUPS

A collection of elements $\{\alpha, \beta, \gamma,..., \sigma,...\}$ comprises a *group* G if the following axioms are satisfied:

1. The product γ of any two elements α and β of the group, which is given by $\gamma = \alpha\beta$, must be a unique element that also belongs to the group.
2. Among the elements of G, there must exist an identity element e which, when multiplied with any element α of the group, leaves the element unchanged: $e\alpha = \alpha e = \alpha$.
3. Each element α of G must have an inverse α^{-1} also belonging to the group, such that $\alpha\alpha^{-1} = \alpha^{-1}\alpha = e$.
4. When three or more elements of G are multiplied, the order of the multiplications is immaterial: $\alpha(\beta\gamma) = (\alpha\beta)\gamma$.

A group where all elements are symmetry operations constitutes a *symmetry group*. Symmetry operations may be *reflections* in planes of symmetry (denoted by σ_l, where l is the plane of symmetry), *rotations* about an axis of symmetry (denoted by C_n, if the angle of rotation is $2\pi/n$), *rotary-reflections* S_n (denoting a rotation through an angle $2\pi/n$, combined with a reflection in the plane perpendicular to the rotation axis), or *inversions i* (which are reflections in the *centre of symmetry*, that is, the one point of a space object that is unmoved by all symmetry operations). Clearly, an inversion is equivalent to a rotary-reflection of angle of rotation π.

In all our considerations, we will assume that all planes of symmetry are vertical, unless otherwise stated; hence, we may denote reflection planes by the symbol v instead of l.

Classification of symmetry groups is usually based on the types of symmetry elements making them up. For example, groups denoted by C_n and C_{nv} all possess a single n-fold axis of rotational symmetry; they are of *orders* n and $2n$, respectively, the order of a group being simply the total number of elements comprising it. Thus, each of these groups possesses n rotation elements (one of which is actually the identity element e), with the C_{nv} family having an additional n reflection elements.

As an example, the simple group C_{1v}, with elements e (identity) and σ_v (reflection in a vertical plane of symmetry), describes the symmetry of any physical configuration with only one plane of symmetry.

Two groups commonly encountered in structural engineering configurations are the C_{2v} and C_{4v} groups. With reference to the xyz Cartesian coordinate system, let us consider two-dimensional configurations situated in the xy plane (assumed to be horizontal), or the projections on the xy plane of three-dimensional configurations whose symmetry properties can all be projected onto the xy plane.

The group C_{2v}, with elements e (identity), C_2 (180° in-plane rotation about the vertical axis through the centre of symmetry), σ_x (reflection in a vertical plane parallel to the xz plane and passing through the centre of symmetry) and σ_y (reflection in a vertical plane parallel to the yz plane and passing through the centre of symmetry), is applicable to configurations with the same symmetry as a rectangle.

The group C_{4v} contains all four elements of the group C_{2v}, and also has the additional elements C_4 (90° *clockwise* rotation about the vertical axis through the centre of symmetry), C_4^{-1} (90° *anticlockwise* rotation about the vertical axis through the centre of symmetry), σ_1 (reflection in a vertical plane inclined at 45° to the xz and yz planes, and passing through the centre of symmetry) and σ_2 (reflection in a vertical plane perpendicular to the vertical plane corresponding to σ_1, and passing through the centre of symmetry); it is thus applicable to configurations with the same symmetry as a square.

6.2 GROUP TABLES AND CLASSES

Multiplication combinations of group elements generate the *group table*, which facilitates the evaluation of products of three or more elements. For illustrative purposes, the group tables for groups C_{1v}, C_{2v} and C_{4v} are presented as Tables 6.1 to 6.3, respectively. In these tables, the name of the group is shown in the top left corner. The order of the multiplication is defined by $\alpha\beta = \gamma$, where the element α is taken from the left side and the element β from the top, their product γ being entered at the intersection of row α and column β.

As expected from the first of the group axioms given earlier, all entries of a group table are elements of the group itself. Groups C_{1v} and C_{2v} are described as *Abelian* because their group tables are symmetrical about the principal diagonal, which implies that group elements multiply commutatively.

Table 6.1 Group table
for group C_{1v}

C_{1v}	e	σ_v
e	e	σ_v
σ_v	σ_v	e

Table 6.2 Group table for group C_{2v}

C_{2v}	e	C_2	σ_x	σ_y
e	e	C_2	σ_x	σ_y
C_2	C_2	e	σ_y	σ_x
σ_x	σ_x	σ_y	e	C_2
σ_y	σ_y	σ_x	C_2	e

Table 6.3 Group table for group C_{4v}

C_{4v}	e	C_4	C_4^{-1}	C_2	σ_x	σ_y	σ_1	σ_2
e	e	C_4	C_4^{-1}	C_2	σ_x	σ_y	σ_1	σ_2
C_4	C_4	C_2	e	C_4^{-1}	σ_1	σ_2	σ_y	σ_x
C_4^{-1}	C_4^{-1}	e	C_2	C_4	σ_2	σ_1	σ_x	σ_y
C_2	C_2	C_4^{-1}	C_4	e	σ_y	σ_x	σ_2	σ_1
σ_x	σ_x	σ_2	σ_1	σ_y	e	C_2	C_4^{-1}	C_4
σ_y	σ_y	σ_1	σ_2	σ_x	C_2	e	C_4	C_4^{-1}
σ_1	σ_1	σ_x	σ_y	σ_2	C_4	C_4^{-1}	e	C_2
σ_2	σ_2	σ_y	σ_x	σ_1	C_4^{-1}	C_4	C_2	e

An element α of a group G is said to be *conjugate* to the element β in the same group if there exists an element ρ in G, such that $\alpha = \rho^{-1}\beta\rho$. The collection of all elements formed by evaluating $\rho^{-1}\beta\rho$ for all ρ in the group is called the *class* of β. The elements of any finite group can be divided into nonoverlapping classes. To establish the classes of a group, it is best to construct the *class table* first, by evaluating $\rho^{-1}\beta\rho$ for all elements ρ and β of the group, ρ being taken from the left side and β from the top, the product $\rho^{-1}\beta\rho$ being entered at the intersection of row ρ and column β.

The class tables for groups C_{1v}, C_{2v} and C_{4v} are given as Tables 6.4 to 6.6, respectively. Collecting into a set the distinct elements appearing in each column of the class table, for all columns of the table (but without

Table 6.4 Class table
for group C_{1v}

C_{1v}	e	σ_v
e	e	σ_v
σ_v	e	σ_v

Table 6.5 Class table for group C_{2v}

C_{2v}	e	C_2	σ_x	σ_y
e	e	C_2	σ_x	σ_y
C_2	e	C_2	σ_x	σ_y
σ_x	e	C_2	σ_x	σ_y
σ_y	e	C_2	σ_x	σ_y

Table 6.6 Class table for group C_{4v}

C_{4v}	e	C_4	C_4^{-1}	C_2	σ_x	σ_y	σ_1	σ_2
e	e	C_4	C_4^{-1}	C_2	σ_x	σ_y	σ_1	σ_2
C_4	e	C_4	C_4^{-1}	C_2	σ_y	σ_x	σ_2	σ_1
C_4^{-1}	e	C_4	C_4^{-1}	C_2	σ_y	σ_x	σ_2	σ_1
C_2	e	C_4	C_4^{-1}	C_2	σ_x	σ_y	σ_1	σ_2
σ_x	e	C_4^{-1}	C_4	C_2	σ_x	σ_y	σ_2	σ_1
σ_y	e	C_4^{-1}	C_4	C_2	σ_x	σ_y	σ_2	σ_1
σ_1	e	C_4^{-1}	C_4	C_2	σ_y	σ_x	σ_1	σ_2
σ_2	e	C_4^{-1}	C_4	C_2	σ_y	σ_x	σ_1	σ_2

repeating any sets), we obtain the classes of groups C_{1v}, C_{2v} and C_{4v} as follows:

Group C_{1v}

$$K_1 = \{e\}; K_2 = \{\sigma_v\}$$

Group C_{2v}

$$K_1 = \{e\}; K_2 = \{C_2\}; K_3 = \{\sigma_x\}; K_4 = \{\sigma_y\}$$

Group C_{4v}

$$K_1 = \{e\}; K_2 = \{C_4, C_4^{-1}\}; K_3 = \{C_2\}; K_4 = \{\sigma_x, \sigma_y\}; K_5 = \{\sigma_1, \sigma_2\}$$

It may be noted that the identity element e always belongs to a class of its own, and that for the Abelian groups C_{1v} and C_{2v}, every element is also in a class of its own.

6.3 REPRESENTATIONS OF SYMMETRY GROUPS

Let a set of symmetry operators $\{\alpha, \beta, \gamma, ..., \sigma,...\}$ in an n-dimensional *vector space* V constitute a group G. Then the set of matrices describing all the symmetry operators of G, with respect to a particular *basis* of the n-dimensional vector space, constitutes an n-dimensional *representation* of G. The *trace* of a matrix (i.e. the sum of the diagonal elements) representing an operator σ is called the *character* of σ, henceforth denoted by $\chi(\sigma)$.

If the basis of V is changed, a new set of matrices, also constituting a representation of the group G, results. Noting that a *basis transformation* does not change the trace of a matrix representing a linear operator in an n-dimensional vector space, this second representation of G possesses the same set of characters as the first representation.

Suppose now that a basis is found with respect to which the matrices of the group representation are expressed as direct sums of submatrices that no further change of basis can reduce to matrices of smaller dimensions. Then sets of these submatrices (a set covering all the symmetry elements of the group) constitute *irreducible representations*. Through this process, the vector space V would have divided into a number of *group-invariant subspaces*, such that none of these subspaces can be divided into further group-invariant subspaces of smaller dimensions.

According to *representation theory* [2], if a group G of order h has k classes, then the number of different irreducible representations is finite and equal to k. Thus, for example, symmetry groups C_{1v}, C_{2v} and C_{4v} have two, four and five different irreducible representations, respectively, implying that the vector space V of problems involving C_{1v}, C_{2v} and C_{4v} symmetry groups can be decomposed into two, four and five group-invariant subspaces, respectively.

6.4 CHARACTER TABLES

In any irreducible representation, all group elements belonging to the same class have the same character, since traces of conjugate elements are equal. Denoting the k different classes of a group G by $K_1, K_2, ..., K_k$, the k different irreducible representations by $R_1, R_2, ..., R_k$, and the character that belongs to the class K_j in the representation R_i by χ_{ij}, characters can be listed in the format depicted in Table 6.7.

Table 6.7 General format for character tables

G	K_1	K_2	·	K_j	·	K_k
R_1	χ_{11}	χ_{12}	·	χ_{1j}	·	χ_{1k}
R_2	χ_{21}	χ_{22}	·	χ_{2j}	·	χ_{2k}
·	·	·	·	·		·
R_i	χ_{i1}	χ_{i2}	·	χ_{ij}	·	χ_{ik}
·	·	·	·	·		·
R_k	χ_{k1}	χ_{k2}	·	χ_{kj}	·	χ_{kk}

Table 6.8 Character table for group C_{1v}

C_{1v}	{e}	$\{\sigma_v\}$
A_1	1	1
A_2	1	−1

Table 6.9 Character table for group C_{2v}

C_{2v}	{e}	$\{C_2\}$	$\{\sigma_x\}$	$\{\sigma_y\}$
A_1	1	1	1	1
A_2	1	1	−1	−1
B_1	1	−1	1	−1
B_2	1	−1	−1	1

Table 6.10 Character table for group C_{4v}

C_{4v}	{e}	$\{C_4, C_4^{-1}\}$	$\{C_2\}$	$\{\sigma_x, \sigma_y\}$	$\{\sigma_1, \sigma_2\}$
A_1	1	1	1	1	1
A_2	1	1	1	−1	−1
B_1	1	−1	1	1	−1
B_2	1	−1	1	−1	1
E	2	0	−2	0	0

Character tables for groups C_{1v}, C_{2v} and C_{4v} are given as Tables 6.8 to 6.10; comprehensive lists of character tables for common symmetry groups are widely available [1–4]. The sets of characters for different irreducible representations are always different. Any two rows of the character table are orthogonal, as are any two columns.

6.5 GROUP ALGEBRA

By introducing addition as a second operation into a group G, the *group algebra L* is created. The group algebra, whose elements are all the linear combinations of group elements, has the properties of a vector space. In the group algebra L, the elements of G constitute a basis of L, since they are linearly independent.

All elements of G that commute with every element of L comprise the *centre of the group algebra*. Class sums (that is, sums of group elements that belong to the same class) form a basis of the centre of the group algebra, and the dimension of the centre equals the number of classes.

6.6 IDEMPOTENTS

Idempotents P_i of the group algebra are its nonzero elements that satisfy the relation $P_i^2 = P_i$. Orthogonal idempotents satisfy the relation $P_i P_j = 0$ if $i \neq j$. They are linearly independent; the sum of orthogonal idempotents is also an idempotent.

Idempotents of the centre of the group algebra are linear combinations of class sums. An idempotent P_i corresponding to the irreducible representation R_i, by operating on vectors of the space V, nullifies every vector that does not belong to the subspace S_i of R_i. Thus, out of all the vectors belonging to the group-invariant subspaces $S_1, S_2, ..., S_k$, the operator P_i selects all vectors belonging to the subspace S_i (which all have a definite symmetry type characteristic of the subspace), and therefore acts as a *projection operator* [2] of the subspace S_i.

In particular, when applied as positional operators upon the positional functions $\phi_1, \phi_2, ..., \phi_n$ of an n degree-of-freedom vibration problem with one degree-of-freedom per position, idempotents generate the *symmetry-adapted positional functions* for their respective subspaces, enabling basis vectors for the various subspaces to be written down.

The orthogonal idempotents of the centre of the group algebra (P_i for subspace S_i, for $i = 1,2,...,k$) can be written down directly from the character table of the group, using the relation [8]

$$P_i = \frac{h_i}{h} \sum_{\sigma} \chi_i(\sigma^{-1})\sigma \tag{6.1}$$

where h is the order of G (that is, the number of elements of G), h_i is the dimension of the ith irreducible representation (given by $h_i = \chi_i(e)$, the first character of the ith row of the character table), χ_i is a character of the ith irreducible representation, σ is an element of G and σ^{-1} its inverse. On the basis of this formula, expressions (6.2) to (6.4) give the idempotents for groups C_{1v}, C_{2v} and C_{4v}, respectively.

Group C_{1v}

$$P_1 = \frac{1}{2}(e + \sigma_v) \tag{6.2a}$$

$$P_2 = \frac{1}{2}(e - \sigma_v) \tag{6.2b}$$

Group C_{2v}

$$P_1 = \frac{1}{4}(e + C_2 + \sigma_x + \sigma_y) \tag{6.3a}$$

$$P_2 = \frac{1}{4}(e + C_2 - \sigma_x - \sigma_y) \tag{6.3b}$$

$$P_3 = \frac{1}{4}(e - C_2 + \sigma_x - \sigma_y) \tag{6.3c}$$

$$P_4 = \frac{1}{4}(e - C_2 - \sigma_x + \sigma_y) \tag{6.3d}$$

Group C_{4v}

$$P_1 = \frac{1}{8}\left(e + C_4 + C_4^{-1} + C_2 + \sigma_x + \sigma_y + \sigma_1 + \sigma_2\right) \tag{6.4a}$$

$$P_2 = \frac{1}{8}\left(e + C_4 + C_4^{-1} + C_2 - \sigma_x - \sigma_y - \sigma_1 - \sigma_2\right) \tag{6.4b}$$

$$P_3 = \frac{1}{8}\left(e - C_4 - C_4^{-1} + C_2 + \sigma_x + \sigma_y - \sigma_1 - \sigma_2\right) \tag{6.4c}$$

$$P_4 = \frac{1}{8}\left(e - C_4 - C_4^{-1} + C_2 - \sigma_x - \sigma_y + \sigma_1 + \sigma_2\right) \tag{6.4d}$$

$$P_5 = \frac{1}{2}(e - C_2) \tag{6.4e}$$

A more detailed treatment of symmetry groups and their representations, as well as of character tables, group algebra and idempotents, may be seen in the books by Hamermesh [2] and Zlokovic [8].

6.7 APPLICATIONS

If a physical system possesses symmetry properties that can be described by a group, decomposition of the vector space V of the system on the basis of the outlined considerations results in a block-diagonal form of the matrix of equations describing the behaviour of the system, allowing separation into independent sets of equations, each corresponding to a particular subspace of V.

In eigenvalue problems, all eigenvalues of the space V of the system can be generated by separately solving k smaller eigenvalue problems, each corresponding to a subspace S_i $(i = 1,2,...,k)$ of the problem. *The eigenvalues yielded from the individual subspaces by the group-theoretic procedure are also the actual eigenvalues of the original problem.* The procedure affords large savings in computational effort in comparison with the conventional approach.

In Chapters 7 to 9, the above concepts are applied to the vibration analysis of symmetric configurations of rectilinear mechanical models [6], plane structural grids [5] and high-tension cable nets [7]. In all cases, the computational advantages of group-theoretic formulations over conventional methods are very evident. Group theory is also valuable in gaining insights into the vibration characteristics of complex structural systems before any calculations are actually performed [9]. Even if symmetry is not evident in the physical system, a symmetric equivalent may exist, permitting group-theoretic techniques to be employed [10].

With numerical implementation in mind, group-theoretic finite-difference and finite-element formulations are presented in Chapters 10 and 11, based on the previous work of the author [11, 12]. A comprehensive review of group-theoretic applications in solid and structural mechanics (including vibration analysis) first appeared in the literature about 10 years ago [13], and was updated more recently [14].

The material in the remaining part of this book represents a relatively new and novel approach to the vibration analysis of structural and mechanical systems. For simplicity, damping will be ignored throughout. However, the considerations are quite general, as long as the requirement for symmetry is preserved.

In some places, the notation that is used in this part of the book differs from that which was employed in Part I. This should not cause any confusion at all, since Part II of the book is almost self-contained. In any case, it is assumed that the reader of Part II is already familiar with the fundamentals of vibration analysis (which were covered in Part I) and would not need to refer to Part I except on very few occasions.

REFERENCES

1. A.P. Cracknell. *Applied Group Theory*. Pergamon Press, Oxford, 1968.
2. M. Hamermesh. *Group Theory and Its Application to Physical Problems*. Pergamon Press, Oxford, 1962.
3. J.W. Leech and D.J. Newman. *How to Use Groups*. Methuen, London, 1969.
4. D. Schonland. *Molecular Symmetry*. Van Nostrand, London, 1965.
5. A. Zingoni and M.N. Pavlovic. On Natural-Frequency Determination of Symmetric Grid-Mass Systems. In *Proceedings of the Fifth International Conference on Recent Advances in Structural Dynamics* (edited by N.S. Ferguson, H.F. Wolfe and C. Mei), Institute of Sound and Vibration Research, Southampton, 1994, 151–163.
6. A. Zingoni. A New Approach for the Vibration Analysis of Symmetric Mechanical Systems. Part 2. One-Dimensional Systems. *International Journal of Engineering Education*, 1996, 12(2), 152–157.
7. A. Zingoni. An Efficient Computational Scheme for the Vibration Analysis of High-Tension Cable Nets. *Journal of Sound and Vibration*, 1996, 189(1), 55–79.
8. G.M. Zlokovic. *Group Theory and G-Vector Spaces in Structural Analysis*. Ellis Horwood, Chichester, 1989.
9. A. Zingoni. On the Symmetries and Vibration Modes of Layered Space Grids. *Engineering Structures*, 2005, 27(4), 629–638.
10. A. Zingoni. On Group-Theoretic Computation of Natural Frequencies for Spring–Mass Dynamic Systems with Rectilinear Motion. *Communications in Numerical Methods in Engineering*, 2008, 24, 973–987.
11. A. Zingoni. A Group-Theoretic Finite-Difference Formulation for Plate Eigenvalue Problems. *Computers and Structures*, 2012, 112/113, 266–282.
12. A. Zingoni. A Group-Theoretic Formulation for Symmetric Finite Elements. *Finite Elements in Analysis and Design*, 2005, 41(6), 615–635.
13. A. Zingoni. Group-Theoretic Applications in Solid and Structural Mechanics: A Review. In *Computational Structures Technology* (edited by B.H.V. Topping and Z. Bittnar), Saxe-Coburg Publications, Stirling, Scotland, 2002, pp. 283–317.
14. A. Zingoni. Group-Theoretic Exploitations of Symmetry in Computational Solid and Structural Mechanics. *International Journal for Numerical Methods in Engineering*, 2009, 79, 253–289.

Chapter 7

Rectilinear models

7.1 INTRODUCTION

Two examples of one-dimensional symmetric systems are considered in turn. These comprise the rotational vibrations of a shaft–disc system and the extensional vibrations of a spring–mass system, the first having an even number of degrees of freedom ($n = 6$) and the other having an odd number ($n = 5$). Both configurations belong to the symmetry group C_{1v}, which was described in Chapter 6.

Using the idempotents for group C_{1v} (see Equation (6.2) of Chapter 6), basis vectors (that is, *symmetry-adapted functions*) spanning the respective subspaces of each of the two problems are deduced. These enable *symmetry-adapted stiffness matrices* (corresponding to the symmetric and antisymmetric subspaces) to be derived, leading to two independent characteristic equations.

If n is even, these equations are both of degree $n/2$; if n is odd, one equation will be of degree $(n+1)/2$, while the other will be of degree $(n-1)/2$. In either case, the computational effort that would be incurred in solving for the eigenvalues is considerably less than were a polynomial equation of full degree n (as yielded by the conventional method) to be solved.

7.2 A SHAFT–DISC TORSIONAL SYSTEM

Let us consider a shaft with both ends fixed, and having six discs attached to it in a symmetrical configuration, as shown in Figure 7.1. The system has six degrees of freedom describing the rotational motions θ_1, θ_2, θ_3, θ_4, θ_5 and θ_6 of discs 1, 2, 3, 4, 5 and 6. The mass moments of inertia (I) of the discs, and the torsional constants (k) of the various shaft intervals, are as shown in the figure.

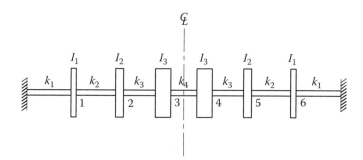

Figure 7.1 Shaft–disc torsional system with $n = 6$.

7.2.1 Symmetry-adapted functions

The symmetry-adapted functions of the two subspaces S_1 and S_2 associated with the group C_{1v} are obtained by applying the respective idempotents (P_1 in the case of subspace S_1, and P_2 in the case of subspace S_2) to the system functions $\phi_1(=\theta_1)$, $\phi_2(=\theta_2)$, $\phi_3(=\theta_3)$, $\phi_4(=\theta_4)$, $\phi_5(=\theta_5)$ and $\phi_6(=\theta_6)$, as follows:

Subspace S_1

$$P_1\phi_1 = \frac{1}{2}(e+C_2)\phi_1 = \frac{1}{2}(\phi_1 + \phi_6) = P_1\phi_6$$

$$P_1\phi_2 = \frac{1}{2}(e+C_2)\phi_2 = \frac{1}{2}(\phi_2 + \phi_5) = P_1\phi_5$$

$$P_1\phi_3 = \frac{1}{2}(e+C_2)\phi_3 = \frac{1}{2}(\phi_3 + \phi_4) = P_1\phi_4$$

Thus, the symmetrical subspace S_1 is three-dimensional (that is, it has three linearly independent vectors). The basis vectors may be taken as follows:

$$\Phi_1 = \phi_1 + \phi_6 \tag{7.1a}$$

$$\Phi_2 = \phi_2 + \phi_5 \tag{7.1b}$$

$$\Phi_3 = \phi_3 + \phi_4 \tag{7.1c}$$

Subspace S_2

$$P_2\phi_1 = \frac{1}{2}(e - C_2)\phi_1 = \frac{1}{2}(\phi_1 - \phi_6) = -P_2\phi_6$$

$$P_2\phi_2 = \frac{1}{2}(e - C_2)\phi_2 = \frac{1}{2}(\phi_2 - \phi_5) = -P_2\phi_5$$

$$P_2\phi_3 = \frac{1}{2}(e - C_2)\phi_3 = \frac{1}{2}(\phi_3 - \phi_4) = -P_2\phi_4$$

Thus, the antisymmetrical subspace S_2 is also three-dimensional. The basis vectors for this subspace may be taken as follows:

$$\Phi_1 = \phi_1 - \phi_6 \tag{7.2a}$$

$$\Phi_2 = \phi_2 - \phi_5 \tag{7.2b}$$

$$\Phi_3 = \phi_3 - \phi_4 \tag{7.2c}$$

7.2.2 Symmetry-adapted stiffness matrices

Let us denote these by K_1 and K_2 for subspaces S_1 and S_2, respectively. These matrices will both be 3×3 in size, since both S_1 and S_2 are three-dimensional.

To generate the symmetry-adapted stiffness matrices K_1 (for the symmetrical subspace) and K_2 (for the antisymmetrical subspace), unit rotations must be applied upon the system (and the resisting torsional moments noted), in accordance with the ϕ coordinates of the respective basis vectors, as elaborated below.

Each of the sketches in Figure 7.2 shows a longitudinal section of the system through the shaft axis, with the shaft itself and the cross sections of its discs merely shown as lines. A dot in the square at one end of a cross section of a disc denotes movement (of this end of the disc cross section) *toward* the observer, while a cross denotes movement *away from* the observer, this convention for tangential motion of the disc edges across the plane of the sketches automatically defining the relative senses of the unit rotations associated with a given basis vector.

For either subspace, Φ_1 has components ϕ_1 and ϕ_6, Φ_2 has components ϕ_2 and ϕ_5, and Φ_3 has components ϕ_3 and ϕ_4 – see expressions (7.1) and (7.2). For the purposes of defining the coefficients of the symmetry-adapted stiffness matrices, disc locations 1 and 6 on the shaft will thus be referred to as the *stations* of Φ_1. Similarly, locations 2 and 5 will be referred to as the stations of Φ_2, while locations 3 and 4 will be referred to as the stations of Φ_3.

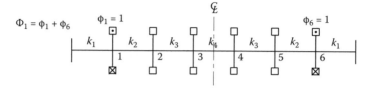

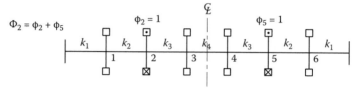

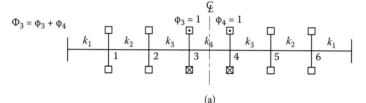

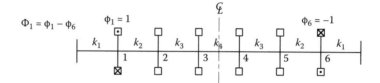

(a)

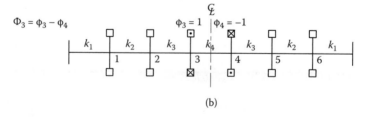

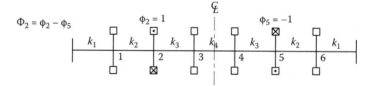

(b)

Figure 7.2 Unit rotations of discs applied in accordance with the coordinates of the basis vectors for the shaft–disc torsional system: (a) subspace S_1, (b) subspace S_2.

The stiffness coefficients for K_1 or K_2 are obtained in very much the same way as in the conventional procedure. However, instead of the six rotation variables $\{\phi_1, \phi_2, \phi_3, \phi_4, \phi_5, \phi_6\}$ of the conventional approach, only three symmetry-adapted rotation functions $\{\Phi_1, \Phi_2, \Phi_3\}$ are now the variables of interest. Also, instead of considering as many as six locations $\{1, 2, 3, 4, 5, 6\}$ in assessing the effects of the unit rotations, the two stations of a given Φ_i ($i = 1, 2, 3$) are treated simultaneously, resulting in only three independent locations at which the effects of unit coordinates of a given rotation vector (Φ_1, Φ_2 or Φ_3) are sought.

Thus, for each subspace, the stiffness coefficient k_{ij} ($i = 1, 2, 3; j = 1, 2, 3$) is the value of the moment at any of the (two) stations of Φ_i due to unit rotations at all the stations of Φ_j, while all the discs other than those at the stations of Φ_j are held at rest (see Figure 7.2). The senses (clockwise or anticlockwise) of the unit rotations applied at the stations of Φ_j are given by the coefficients (+1 or −1) of the components of Φ_j, as appear in Equation (7.1) (in the case of subspace S_1) or Equation (7.2) (in the case of subspace S_2).

In terms of the torsional constants shown in Figure 7.1, the results are as follows:

Subspace S_1

$$k_{11} = k_1 + k_2; \ k_{12} = -k_2; \ k_{13} = 0$$

$$k_{21} = -k_2; \ k_{22} = k_2 + k_3; \ k_{23} = -k_3$$

$$k_{31} = 0; \ k_{32} = -k_3; \ k_{33} = k_3$$

that is,

$$K_1 = \begin{bmatrix} (k_1 + k_2) & -k_2 & 0 \\ -k_2 & (k_2 + k_3) & -k_3 \\ 0 & -k_3 & k_3 \end{bmatrix} \tag{7.3}$$

Subspace S_2

$$k_{11} = k_1 + k_2; \ k_{12} = -k_2; \ k_{13} = 0$$

$$k_{21} = -k_2; \ k_{22} = k_2 + k_3; \ k_{23} = -k_3$$

$$k_{31} = 0; \ k_{32} = -k_3; \ k_{33} = k_3 + 2k_4$$

that is,

$$
K_2 = \begin{bmatrix} (k_1 + k_2) & -k_2 & 0 \\ -k_2 & (k_2 + k_3) & -k_3 \\ 0 & -k_3 & (k_3 + 2k_4) \end{bmatrix}
\tag{7.4}
$$

7.2.3 Eigenvalues

These are obtained from the vanishing condition (refer to Part I, if necessary)

$$
|K - \lambda M| = 0
\tag{7.5}
$$

where $\lambda = \omega^2$, and ω is a natural circular frequency of the system. The subspaces are considered independently of each other, so that for subspace S_1, the stiffness matrix K assumes the form K_1, while for subspace S_2, K assumes the form K_2. For either subspace, the nonzero elements of the 3×3 diagonal mass matrix M are simply the mass moments of inertia occurring at (1) any of the two stations of Φ_1, (2) any of the two stations of Φ_2 and (3) any of the two stations of Φ_3, that is,

$$
M_1 = M_2 = \begin{bmatrix} I_1 & 0 & 0 \\ 0 & I_2 & 0 \\ 0 & 0 & I_3 \end{bmatrix}
\tag{7.6}
$$

Applying condition (7.5) to the two subspaces in turn, we obtain the following equations:

Subspace S_1

$$
\begin{vmatrix} \left(\dfrac{k_1 + k_2}{I_1} - \lambda\right) & -\dfrac{k_2}{I_1} & 0 \\ -\dfrac{k_2}{I_2} & \left(\dfrac{k_2 + k_3}{I_2} - \lambda\right) & -\dfrac{k_3}{I_2} \\ 0 & -\dfrac{k_3}{I_3} & \left(\dfrac{k_3}{I_3} - \lambda\right) \end{vmatrix} = 0
\tag{7.7}
$$

Subspace S_2

$$\begin{vmatrix} \left(\dfrac{k_1 + k_2}{I_1} - \lambda\right) & -\dfrac{k_2}{I_1} & 0 \\[3mm] -\dfrac{k_2}{I_2} & \left(\dfrac{k_2 + k_3}{I_2} - \lambda\right) & -\dfrac{k_3}{I_2} \\[3mm] 0 & -\dfrac{k_3}{I_3} & \left(\dfrac{k_3 + 2k_4}{I_3} - \lambda\right) \end{vmatrix} = 0 \qquad (7.8)$$

Expanding the determinant in each of the above equations leads to two independent cubic equations in λ, which, upon solving separately, yield the six required eigenvalues λ_1, λ_2, λ_3, λ_4, λ_5 and λ_6. This represents a considerable simplification in the determination of the eigenvalues, in comparison with the conventional approach that would require the evaluation of a full 6×6 determinant – not a trivial task – and the solution of an ensuing polynomial equation of sixth degree.

7.3 A SPRING–MASS EXTENSIONAL SYSTEM

Our second example involves a spring–mass system with five degrees of freedom, describing the extensional motions x_1, x_2, x_3, x_4 and x_5 of masses m_1 (at locations 1 and 5), m_2 (at locations 2 and 4) and m_3 (at location 3), as shown in Figure 7.3. The spring constants (k) of the various intervals of the system are also shown in the figure. It is clear that the stationary configuration of the system is symmetrical about a perpendicular plane through location 3, and thus belongs to the group C_{1v}.

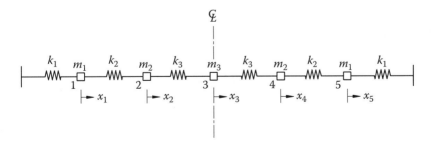

Figure 7.3 Spring–mass extensional system with $n = 5$.

7.3.1 Symmetry-adapted functions

Applying the respective idempotents for subspaces S_1 and S_2 of group C_{1v} to the system functions $\phi_1(=x_1)$, $\phi_2(=x_2)$, $\phi_3(=x_3)$, $\phi_4(=x_4)$ and $\phi_5(=x_5)$, we obtain the symmetry-adapted functions for the two subspaces as follows:

Subspace S_1

$$P_1\phi_1 = \frac{1}{2}(e+C_2)\phi_1 = \frac{1}{2}(\phi_1+\phi_5) = P_1\phi_5$$

$$P_1\phi_2 = \frac{1}{2}(e+C_2)\phi_2 = \frac{1}{2}(\phi_2+\phi_4) = P_1\phi_4$$

$$P_1\phi_3 = \frac{1}{2}(e+C_2)\phi_3 = \frac{1}{2}(\phi_3+\phi_3) = \phi_3$$

Thus, the symmetrical subspace S_1 is three-dimensional. The basis vectors may be taken as

$$\Phi_1 = \phi_1 + \phi_5 \tag{7.9a}$$

$$\Phi_2 = \phi_2 + \phi_4 \tag{7.9b}$$

$$\Phi_3 = \phi_3 \tag{7.9c}$$

Subspace S_2

$$P_2\phi_1 = \frac{1}{2}(e-C_2)\phi_1 = \frac{1}{2}(\phi_1-\phi_5) = -P_2\phi_5$$

$$P_2\phi_2 = \frac{1}{2}(e-C_2)\phi_2 = \frac{1}{2}(\phi_2-\phi_4) = -P_2\phi_4$$

$$P_2\phi_3 = \frac{1}{2}(e-C_2)\phi_3 = \frac{1}{2}(\phi_3-\phi_3) = 0$$

Thus, the antisymmetrical subspace S_2 is two-dimensional. The two basis vectors may be taken as follows:

$$\Phi_1 = \phi_1 - \phi_5 \tag{7.10a}$$

$$\Phi_2 = \phi_2 - \phi_4 \tag{7.10b}$$

7.3.2 Symmetry-adapted stiffness matrices

For each subspace, these are obtained as explained for the shaft–disc example, with unit displacements now taking the place of unit rotations, and axial forces now taking the place of moments. In working out the stiffness coefficients, the unit displacements of the masses are applied in accordance with the coordinates of the basis vectors (which are given by Equations (7.9) and (7.10)), as illustrated in Figure 7.4. Thus, for a given subspace of dimension r, the stiffness coefficient k_{ij} $(i = 1,...,r; j = 1,...,r)$ is the force acting on the mass located at any of the stations of Φ_i as a result of unit displacements applied at all the stations of Φ_j, while all the masses other than those at the stations of Φ_j are held at rest. The results are as follows:

Subspace S_1

$$k_{11} = k_1 + k_2;\ k_{12} = -k_2;\ k_{13} = 0$$

$$k_{21} = -k_2;\ k_{22} = k_2 + k_3;\ k_{23} = -k_3$$

$$k_{31} = 0;\ k_{32} = -2k_3;\ k_{33} = 2k_3$$

that is,

$$K_1 = \begin{bmatrix} (k_1 + k_2) & -k_2 & 0 \\ -k_2 & (k_2 + k_3) & -k_3 \\ 0 & -2k_3 & 2k_3 \end{bmatrix} \tag{7.11}$$

Subspace S_2

$$k_{11} = k_1 + k_2;\ k_{12} = -k_2$$

$$k_{21} = -k_2;\ k_{22} = k_2 + k_3$$

that is,

$$K_2 = \begin{bmatrix} (k_1 + k_2) & -k_2 \\ -k_2 & (k_2 + k_3) \end{bmatrix} \tag{7.12}$$

$\Phi_1 = \phi_1 + \phi_5$

$\Phi_2 = \phi_2 + \phi_4$

$\Phi_3 = \phi_3$

(a)

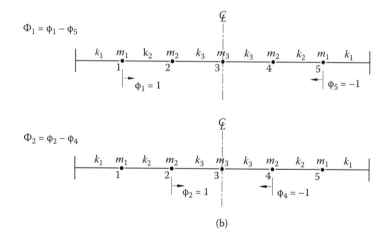

$\Phi_1 = \phi_1 - \phi_5$

$\Phi_2 = \phi_2 - \phi_4$

(b)

Figure 7.4 Unit displacements of masses applied in accordance with the coordinates of the basis vectors for the spring–mass extensional system: (a) subspace S_1, (b) subspace S_2.

7.3.3 Eigenvalues

Applying condition (7.5), with K_1 and K_2 in place of K for subspaces S_1 and S_2, respectively, and the mass matrices

$$M_1 = \begin{bmatrix} m_1 & 0 & 0 \\ 0 & m_2 & 0 \\ 0 & 0 & m_3 \end{bmatrix} \tag{7.13}$$

$$M_2 = \begin{bmatrix} m_1 & 0 \\ 0 & m_2 \end{bmatrix} \tag{7.14}$$

in place of M for subspaces S_1 and S_2, respectively, the following equations are obtained:

Subspace S_1

$$\begin{vmatrix} \left(\dfrac{k_1 + k_2}{m_1} - \lambda \right) & -\dfrac{k_2}{m_1} & 0 \\[2ex] -\dfrac{k_2}{m_2} & \left(\dfrac{k_2 + k_3}{m_2} - \lambda \right) & -\dfrac{k_3}{m_2} \\[2ex] 0 & -\dfrac{2k_3}{m_3} & \left(\dfrac{2k_3}{m_3} - \lambda \right) \end{vmatrix} = 0 \tag{7.15}$$

Subspace S_2

$$\begin{vmatrix} \left(\dfrac{k_1 + k_2}{m_1} - \lambda \right) & -\dfrac{k_2}{m_1} \\[2ex] -\dfrac{k_2}{m_2} & \left(\dfrac{k_2 + k_3}{m_2} - \lambda \right) \end{vmatrix} = 0 \tag{7.16}$$

Expanding the determinant in each of the above equations results in two independent equations in λ, one a cubic and the other a quadratic; upon solving these separately, the five required eigenvalues ensue. The conventional method would have required the evaluation of a 5×5 determinant for the full system, and the solution of the resulting fifth-degree polynomial equation.

7.4 CONCLUSIONS

The application of symmetry group C_{1v} to two examples of one-dimensional (rectilinear) systems has been illustrated. These have comprised a shaft–disc torsional system with an even number of degrees of freedom and a spring–mass extensional system with an odd number of degrees of freedom, both configurations having a single plane of symmetry.

Denoting the total number of degrees of freedom of any C_{1v} system by n, group-theoretic decomposition of the problem into the constituent symmetrical and antisymmetrical portions results in two independent polynomial equations each of degree $n/2$ if n is even; if n is odd, the symmetrical subspace yields a polynomial equation of degree $(n + 1)/2$, while the antisymmetrical subspace yields a polynomial of degree $(n - 1)/2$.

In either case, the computational effort involved in solving for the n eigenvalues of the problem on the basis of these two mutually independent subspace polynomials is only a fraction of that which would be incurred were a polynomial equation of full degree n (as yielded by the conventional method) to be tackled.

Chapter 8

Plane structural grids

8.1 INTRODUCTION

Grid-mass models are often used to simulate the dynamic response of real plane-grid structures, where grid members may be assumed to be weightless, with all their mass lumped at discrete points on the grid; usually, the member intersections provide, with reasonable accuracy, convenient lumping sites for the distributed mass of the individual members. Such approximate models are particularly appropriate for rectangular and square grids consisting of two mutually perpendicular families of evenly spaced parallel members.

Quite often elastic grids actually carry at the member intersections concentrated masses whose magnitudes are considerably greater than the sum of the masses of flexural members framing into the respective intersections, as may occur in roof systems with suspended loads, and industrial platforms with plant installations. Clearly, the lumped mass model becomes even more appropriate in such instances.

In this chapter, we illustrate a group-theoretic approach for simplifying the determination of natural frequencies for the small transverse vibrations of symmetric grid-mass systems, by reference to two examples, namely, a rectangular configuration and a square configuration. Although the illustration has been limited to only two symmetry groups, the proposed method is, of course, quite general, and applicable whenever a physical system exhibits symmetry properties. The method can readily be applied to plane grids whose members do not intersect at right angles, as long as the overall configuration conforms to a specific symmetry group.

We assume that the transverse deflections of the grid at all its n nodes (that is, member intersections), due to a unit transverse static force applied at any *one* of these n nodes, are known, or can be worked out using the conventional methods of structural analysis. We then make use of group-theoretic concepts to linearly combine these *basic flexibility coefficients* into *symmetry-adapted flexibility coefficients*, which make up the symmetry-adapted flexibility matrices. Thus, as in Chapter 7, the original problem (in an n-dimensional vector space) is decomposed into a series of k

independent smaller problems, which lie in subspaces of dimensions much smaller than n, leading to significant reductions in computational effort.

As already indicated, the amplitudes of the transverse displacements of the masses during vibration are assumed to be small, so that the system response is essentially linear, and superposition is applicable. Torsional effects are presently ignored, though this does not diminish the generality of the solution approach. Also, the grid members will be assumed to have the same flexural rigidity (EI) throughout, for simplicity.

8.2 RECTANGULAR CONFIGURATIONS

Figure 8.1 shows a very simple eight-noded example of a grid-mass system whose configuration belongs to the rectangular group C_{2v}. The adopted node numbering is shown in Figure 8.1(a), while the assignment of concentrated masses to the nodes of the grid is shown in Figure 8.1(b), such an allocation conforming to the C_{2v} symmetry of the structural configuration. The centre of symmetry is labelled O.

8.2.1 Symmetry-adapted functions

Applying each idempotent P_k $(k = 1,2,3,4)$ – these were given in Chapter 6 as Equation (6.3) – in turn to all eight positional functions $\phi_1, \phi_2,..., \phi_8$ shown

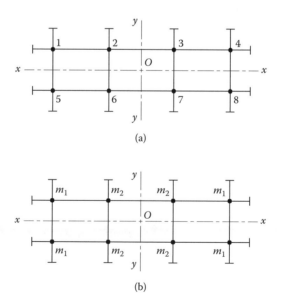

(a)

(b)

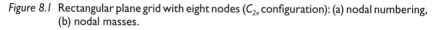

Figure 8.1 Rectangular plane grid with eight nodes (C_{2v} configuration): (a) nodal numbering, (b) nodal masses.

in Figure 8.1(a) yields the symmetry-adapted functions for each subspace S_k (k = 1,2,3,4), from which the independent basis vectors Φ_i (i = 1,2,...,r) spanning an r-dimensional subspace may be selected, as follows:

Subspace S_1

$$P_1\phi_1 = (1/4)(e + C_2 + \sigma_x + \sigma_y)\phi_1$$

$$= (1/4)(\phi_1 + \phi_8 + \phi_5 + \phi_4) = P_1\phi_4 = P_1\phi_5 = P_1\phi_8$$

$$P_1\phi_2 = (1/4)(e + C_2 + \sigma_x + \sigma_y)\phi_2$$

$$= (1/4)(\phi_2 + \phi_7 + \phi_6 + \phi_3) = P_1\phi_3 = P_1\phi_6 = P_1\phi_7$$

$$\Phi_1 = \phi_1 + \phi_4 + \phi_5 + \phi_8 \tag{8.1a}$$

$$\Phi_2 = \phi_2 + \phi_3 + \phi_5 + \phi_7 \tag{8.1b}$$

Subspace S_2

$$P_2\phi_1 = (1/4)(e + C_2 - \sigma_x - \sigma_y)\phi_1$$

$$= (1/4)(\phi_1 + \phi_8 - \phi_5 - \phi_4) = P_2\phi_8 = -P_2\phi_4 = -P_2\phi_5$$

$$P_2\phi_2 = (1/4)(e + C_2 - \sigma_x - \sigma_y)\phi_2$$

$$= (1/4)(\phi_2 + \phi_7 - \phi_6 - \phi_3) = P_2\phi_7 = -P_2\phi_3 = -P_2\phi_6$$

$$\Phi_1 = \phi_1 - \phi_4 - \phi_5 + \phi_8 \tag{8.2a}$$

$$\Phi_2 = \phi_2 - \phi_3 - \phi_6 + \phi_7 \tag{8.2b}$$

Subspace S_3

$$P_3\phi_1 = (1/4)(e - C_2 + \sigma_x - \sigma_y)\phi_1$$

$$= (1/4)(\phi_1 - \phi_8 + \phi_5 - \phi_4) = P_3\phi_5 = -P_3\phi_4 = -P_3\phi_8$$

$$P_3\phi_2 = (1/4)(e - C_2 + \sigma_x - \sigma_y)\phi_2$$

$$= (1/4)(\phi_2 - \phi_7 + \phi_6 - \phi_3) = P_3\phi_6 = -P_3\phi_3 = -P_3\phi_7$$

$$\Phi_1 = \phi_1 - \phi_4 + \phi_5 - \phi_8 \tag{8.3a}$$

$$\Phi_2 = \phi_2 - \phi_3 + \phi_6 - \phi_7 \tag{8.3b}$$

Subspace S_4

$$P_4\phi_1 = (1/4)(e - C_2 - \sigma_x + \sigma_y)\phi_1$$

$$= (1/4)(\phi_1 - \phi_8 - \phi_5 + \phi_4) = P_4\phi_4 = -P_4\phi_5 = -P_4\phi_8$$

$$P_4\phi_2 = (1/4)(e - C_2 - \sigma_x + \sigma_y)\phi_2$$

$$= (1/4)(\phi_2 - \phi_7 - \phi_6 + \phi_3) = P_4\phi_3 = -P_4\phi_6 = -P_4\phi_7$$

$$\Phi_1 = \phi_1 + \phi_4 - \phi_5 - \phi_8 \tag{8.4a}$$

$$\Phi_2 = \phi_2 + \phi_3 - \phi_6 - \phi_7 \tag{8.4b}$$

8.2.2 Symmetry-adapted flexibility coefficients

These are obtained by applying on the grid, for each subspace in turn, unit vertical forces in accordance with the coordinates of the respective basis vectors appearing in Equations (8.1) to (8.4), as illustrated in Figure 8.2.

Now, the conventional flexibility matrix for the configuration shown in Figure 8.1(a) may be written down as f_{ij} ($i = 1,2,...,8$; $j = 1,2,...,8$), where, following usual convention, f_{ij} is the vertical deflection at node i due to a unit vertical force applied at node j. From conventional considerations of the symmetry of the structural configuration in Figure 8.1(a), and with the knowledge that stiffness matrices and flexibility matrices of structures are always symmetrical (a consequence of the reciprocal theorem of elasticity), we may easily show that there are only 12 distinct values of the flexibility coefficients for the structure of Figure 8.1(a), so that the conventional flexibility matrix $\mathbf{A}$ may be assumed to be of the general form

$$\mathbf{A} = \begin{bmatrix}
a_1 & a_2 & a_3 & a_4 & a_5 & a_6 & a_7 & a_8 \\
a_2 & a_9 & a_{10} & a_3 & a_6 & a_{11} & a_{12} & a_7 \\
a_3 & a_{10} & a_9 & a_2 & a_7 & a_{12} & a_{11} & a_6 \\
a_4 & a_3 & a_2 & a_1 & a_8 & a_7 & a_6 & a_5 \\
a_5 & a_6 & a_7 & a_8 & a_1 & a_2 & a_3 & a_4 \\
a_6 & a_{11} & a_{12} & a_7 & a_2 & a_9 & a_{10} & a_3 \\
a_7 & a_{12} & a_{11} & a_6 & a_3 & a_{10} & a_9 & a_2 \\
a_8 & a_7 & a_6 & a_5 & a_4 & a_3 & a_2 & a_1
\end{bmatrix} \tag{8.5}$$

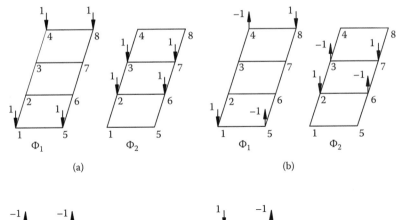

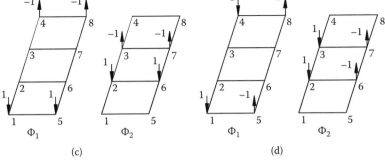

Figure 8.2 Unit vertical forces applied in accordance with the coordinates of the basis vectors, for the rectangular plane grid: (a) subspace S_1, (b) subspace S_2, (c) subspace S_3, (d) subspace S_4.

In the group-theoretic formulation of the problem, let us consider any subspace of dimension r (that is, the subspace is spanned by r independent basis vectors). We will define the *symmetry-adapted flexibility coefficient* b_{ij} ($i = 1,2,...,r$; $j = 1,2,...,r$) as the vertical displacement at any of the nodes of the basis vector Φ_i, due to unit vertical forces applied at all the nodes of the basis vector Φ_j. For the present example, $r = 2$ for all four subspaces, so that the symmetry-adapted flexibility matrices all take the form

$$\mathbf{B} = \begin{bmatrix} b_{11} & b_{12} \\ b_{21} & b_{22} \end{bmatrix} \tag{8.6}$$

For each subspace, the coefficients of $\mathbf{B}$ are obtained by superimposing the appropriate values of the conventional flexibility coefficients appearing in Equation (8.5), in accordance with the combinations of unit vertical forces of Φ_j appearing in Figure 8.2, as follows:

Subspace S_1

$$b_{11} = a_1 + a_4 + a_5 + a_8 \tag{8.7a}$$

$$b_{12} = b_{21} = a_2 + a_3 + a_6 + a_7 \tag{8.7b}$$

$$b_{22} = a_9 + a_{10} + a_{11} + a_{12} \tag{8.7c}$$

Subspace S_2

$$b_{11} = a_1 - a_4 - a_5 + a_8 \tag{8.8a}$$

$$b_{12} = b_{21} = a_2 - a_3 - a_6 + a_7 \tag{8.8b}$$

$$b_{22} = a_9 - a_{10} - a_{11} + a_{12} \tag{8.8c}$$

Subspace S_3

$$b_{11} = a_1 - a_4 + a_5 - a_8 \tag{8.9a}$$

$$b_{12} = b_{21} = a_2 - a_3 + a_6 - a_7 \tag{8.9b}$$

$$b_{22} = a_9 - a_{10} + a_{11} - a_{12} \tag{8.9c}$$

Subspace S_4

$$b_{11} = a_1 + a_4 - a_5 - a_8 \tag{8.10a}$$

$$b_{12} = b_{21} = a_2 + a_3 - a_6 - a_7 \tag{8.10b}$$

$$b_{22} = a_9 + a_{10} - a_{11} - a_{12} \tag{8.10c}$$

8.2.3 Eigenvalues

Noting that the diagonal mass matrix $\mathbf{M}$ for a given subspace consists of nonzero diagonal elements m_{ii} ($i = 1,2,...,r$), which are the values of the mass at each of the nodes of the basis vector Φ_j, we observe, by reference to Figure 8.2 in conjunction with Figure 8.1(b), that all four subspaces have the same mass matrix, namely,

$$\mathbf{M} = \begin{bmatrix} m_1 & 0 \\ 0 & m_2 \end{bmatrix} \tag{8.11}$$

For each subspace, the characteristic equation is given by

$$\left| B - \lambda M^{-1} \right| = 0 \tag{8.12}$$

where $\lambda = 1/\omega^2$, and ω is a natural circular frequency of the system. The above equation is essentially the same as Equation (3.42) of Chapter 3. To see that these two equations are similar, we simply divide Equation (3.42) by λM, rearrange the terms, replace the conventional flexibility matrix D with the symmetry-adapted flexibility matrix B of present considerations, and also replace λ with $1/\lambda$ (since in Part I, λ was defined as $\lambda = \omega^2$, whereas here, we have chosen to define it as $\lambda = 1/\omega^2$). The determinant in Equation (8.12), upon writing out the interior matrices in full, becomes

$$\begin{vmatrix} b_{11} - (\lambda / m_1) & b_{12} \\ b_{21} & b_{22} - (\lambda / m_2) \end{vmatrix} = 0 \tag{8.13}$$

leading to the characteristic equation

$$\lambda^2 - \lambda \left(m_1 b_{11} + m_2 b_{22} \right) + \left(b_{11} b_{22} - b_{12}^2 \right) m_1 m_2 = 0 \tag{8.14}$$

with roots

$$\lambda = \frac{1}{2} \left\{ \left(m_1 b_{11} + m_2 b_{22} \right) \pm \left\{ \left(m_1 b_{11} + m_2 b_{22} \right)^2 - 4 \left(b_{11} b_{22} - b_{12}^2 \right) m_1 m_2 \right\}^{\frac{1}{2}} \right\} \tag{8.15}$$

While the values m_1 and m_2 are, of course, the same for all four subspaces, the values of the coefficients b_{11}, b_{12} and b_{22} differ between the various subspaces, as is clear by reference to Equations (8.7) to (8.10). Expression (8.15) actually represents all of the required eight eigenvalues of the structural system. Note that this is a closed-form solution, since b_{11}, b_{12} and b_{22} are all known for each subspace (their values being given by Equations (8.7) to (8.10)).

Thus, by using group theory, we only needed to solve one quadratic equation in λ, yielding all eight eigenvalues of the problem, instead of solving an eighth-degree polynomial equation as would be yielded by the conventional approach. Clearly, this particular simplification represents a phenomenal reduction in computational effort.

More typically, however, symmetric rectangular grids with a large number n of degrees of freedom (this would necessitate the solution of a polynomial equation of degree n by conventional analysis) would require the solution of four *independent* polynomial equations of degree $n/4$ (or an adjacent integer if $n/4$ is not an integer) by the present approach. In all cases, our approach would provide substantial gains in computational speed or reductions in computer memory requirements.

8.3 SQUARE CONFIGURATIONS

The 16-noded configuration shown in Figure 8.3 belongs to the symmetry group C_{4v}. Figure 8.3(a) shows the assumed numbering for nodes, while Figure 8.3(b) shows the distribution of nodal masses, which also conforms to the C_{4v} configuration of the grid members. The centre of symmetry is denoted by O.

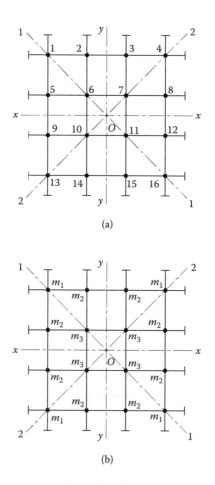

(a)

(b)

Figure 8.3 Square plane grid with 16 nodes (C_{4v} configuration): (a) nodal numbering, (b) nodal masses.

8.3.1 Symmetry-adapted functions

Applying each idempotent P_k (k = 1,2,...,5) – these are given by Equations (6.4) of Chapter 6 – in turn to all 16 positional functions $\phi_1, \phi_2,..., \phi_{16}$ of Figure 8.3(a) yields the symmetry-adapted functions for the corresponding subspace S_k (k = 1,2,...,5), from which the independent basis vectors Φ_i (i = 1,2,...,r) spanning the r-dimensional subspace may be selected.

The eight-dimensional subspace S_5 has been further decomposed into two mutually independent four-dimensional subspaces S_{51} and S_{52} spanned by new basis vectors formed by linearly combining the basis vectors of S_5 in such a way as to form two orthogonal sets. A more direct way for achieving this decomposition will be shown in Chapter 9.

The selected basis vectors for the various subspaces are as follows:

Subspace S_1

$$\Phi_1 = \phi_1 + \phi_4 + \phi_{13} + \phi_{16} \tag{8.16a}$$

$$\Phi_2 = \phi_2 + \phi_3 + \phi_5 + \phi_8 + \phi_9 + \phi_{12} + \phi_{14} + \phi_{15} \tag{8.16b}$$

$$\Phi_3 = \phi_6 + \phi_7 + \phi_{10} + \phi_{11} \tag{8.16c}$$

Subspace S_2

$$\Phi_1 = \phi_2 - \phi_3 - \phi_5 + \phi_8 + \phi_9 - \phi_{12} - \phi_{14} + \phi_{15} \tag{8.17}$$

Subspace S_3

$$\Phi_1 = \phi_2 + \phi_3 - \phi_5 - \phi_8 - \phi_9 - \phi_{12} + \phi_{14} + \phi_{15} \tag{8.18}$$

Subspace S_4

$$\Phi_1 = \phi_1 - \phi_4 - \phi_{13} + \phi_{16} \tag{8.19a}$$

$$\Phi_2 = \phi_2 - \phi_3 + \phi_5 - \phi_8 - \phi_9 + \phi_{12} - \phi_{14} + \phi_{15} \tag{8.19b}$$

$$\Phi_3 = \phi_6 - \phi_7 - \phi_{10} + \phi_{11} \tag{8.19c}$$

Subspace S_{51}

$$\Phi_1 = \phi_1 - \phi_{16} \tag{8.20a}$$

$$\Phi_2 = \phi_6 - \phi_{11} \tag{8.20b}$$

$$\Phi_3 = \phi_2 + \phi_5 - \phi_{12} - \phi_{15} \tag{8.20c}$$

$$\Phi_4 = \phi_3 - \phi_8 + \phi_9 - \phi_{14} \tag{8.20d}$$

Subspace S_{52}

$$\Phi_1 = \phi_4 - \phi_{13} \tag{8.21a}$$

$$\Phi_2 = \phi_7 - \phi_{10} \tag{8.21b}$$

$$\Phi_3 = \phi_3 + \phi_8 - \phi_9 - \phi_{14} \tag{8.21c}$$

$$\Phi_4 = \phi_2 - \phi_5 + \phi_{12} - \phi_{15} \tag{8.21d}$$

8.3.2 Symmetry-adapted flexibility coefficients

These are obtained by applying on the grid, for each subspace in turn, unit vertical forces in accordance with the coordinates of the respective basis vectors appearing in Equations (8.16) to (8.21), as shown in Figure 8.4.

The conventional flexibility matrix for the configuration of Figure 8.3(a) consists of elements f_{ij} $(i = 1,2,...,16; j = 1,2,...,16)$, where as usual f_{ij} is the vertical deflection at node i due to a unit vertical force applied at node j.

From conventional considerations of the symmetry of the structural configuration in Figure 8.3(a), and using the fact that the flexibility matrix is always symmetrical, we may easily show that there are only 24 distinct values of the flexibility coefficients for the structure of Figure 8.3(a), so that the conventional flexibility matrix $\mathbf{A}$ may be assumed to be of the general form

$$
\mathbf{A} = \begin{bmatrix}
a_1 & a_2 & a_3 & a_4 & a_2 & a_5 & a_6 & a_7 & a_3 & a_6 & a_8 & a_9 & a_4 & a_7 & a_9 & a_{10} \\
a_2 & a_{11} & a_{12} & a_3 & a_{13} & a_{14} & a_{15} & a_{16} & a_{16} & a_{17} & a_{18} & a_{19} & a_7 & a_{20} & a_{21} & a_9 \\
a_3 & a_{12} & a_{11} & a_2 & a_{16} & a_{15} & a_{14} & a_{13} & a_{19} & a_{18} & a_{17} & a_{16} & a_9 & a_{21} & a_{20} & a_7 \\
a_4 & a_3 & a_2 & a_1 & a_7 & a_6 & a_5 & a_2 & a_9 & a_8 & a_6 & a_3 & a_{10} & a_9 & a_7 & a_4 \\
a_2 & a_{13} & a_{16} & a_7 & a_{11} & a_{14} & a_{17} & a_{20} & a_{12} & a_{15} & a_{18} & a_{21} & a_3 & a_{16} & a_{19} & a_9 \\
a_5 & a_{14} & a_{15} & a_6 & a_{14} & a_{22} & a_{23} & a_{17} & a_{15} & a_{23} & a_{24} & a_{18} & a_6 & a_{17} & a_{18} & a_8 \\
a_6 & a_{15} & a_{14} & a_5 & a_{17} & a_{23} & a_{22} & a_{14} & a_{18} & a_{24} & a_{23} & a_{15} & a_8 & a_{18} & a_{17} & a_6 \\
a_7 & a_{16} & a_{13} & a_2 & a_{20} & a_{17} & a_{14} & a_{11} & a_{21} & a_{18} & a_{15} & a_{12} & a_9 & a_{19} & a_{16} & a_3 \\
a_3 & a_{16} & a_{19} & a_9 & a_{12} & a_{15} & a_{18} & a_{21} & a_{11} & a_{14} & a_{17} & a_{20} & a_2 & a_{13} & a_{16} & a_7 \\
a_6 & a_{17} & a_{18} & a_8 & a_{15} & a_{23} & a_{24} & a_{18} & a_{14} & a_{22} & a_{23} & a_{17} & a_5 & a_{14} & a_{15} & a_6 \\
a_8 & a_{18} & a_{17} & a_6 & a_{18} & a_{24} & a_{23} & a_{15} & a_{17} & a_{23} & a_{22} & a_{14} & a_6 & a_{15} & a_{14} & a_5 \\
a_9 & a_{19} & a_{16} & a_3 & a_{21} & a_{18} & a_{15} & a_{12} & a_{20} & a_{17} & a_{14} & a_{11} & a_7 & a_{16} & a_{13} & a_2 \\
a_4 & a_7 & a_9 & a_{10} & a_3 & a_6 & a_8 & a_9 & a_2 & a_5 & a_6 & a_7 & a_1 & a_2 & a_3 & a_4 \\
a_7 & a_{20} & a_{21} & a_9 & a_{16} & a_{17} & a_{18} & a_{19} & a_{13} & a_{14} & a_{15} & a_{16} & a_2 & a_{11} & a_{12} & a_3 \\
a_9 & a_{21} & a_{20} & a_7 & a_{19} & a_{18} & a_{17} & a_{16} & a_{16} & a_{15} & a_{14} & a_{13} & a_3 & a_{12} & a_{11} & a_2 \\
a_{10} & a_9 & a_7 & a_4 & a_9 & a_8 & a_6 & a_3 & a_7 & a_6 & a_5 & a_2 & a_4 & a_3 & a_2 & a_1
\end{bmatrix} \tag{8.22}
$$

As with the example of Section 8.2, the coefficients b_{ij} $(i = 1,2,...,r; j = 1,2,...,r)$ of the symmetry-adapted flexibility matrix $\mathbf{B}$ for an r-dimensional subspace are obtained by superimposing the appropriate values of the basic flexibility coefficients appearing in Equation (8.22), in accordance with the

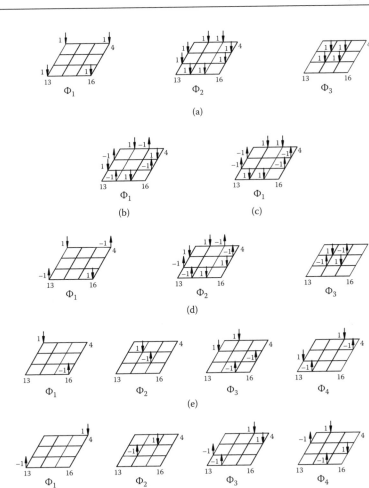

Figure 8.4 Unit vertical forces applied in accordance with the coordinates of the basis vectors, for the square plane grid: (a) subspace S_1, (b) subspace S_2, (c) subspace S_3, (d) subspace S_4, (e) subspace S_{51}, (f) subspace S_{52}.

combinations of unit vertical forces of Φ_j plotted in Figure 8.4. The results for the symmetry-adapted flexibility matrices are as follows:

Subspaces S_1 and S_4

$$\mathbf{B} = \begin{bmatrix} b_{11} & b_{12} & b_{13} \\ b_{21} & b_{22} & b_{23} \\ b_{31} & b_{32} & b_{33} \end{bmatrix}$$

(8.23)

where

$$b_{11} = a_1 \pm a_4 + a_{10} \qquad (8.24a)$$

$$b_{12} = b_{21} = 2a_2 \pm 2a_3 \pm 2a_7 + 2a_9 \qquad (8.24b)$$

$$b_{13} = a_5 \pm 2a_6 + a_8 \qquad (8.24c)$$

$$b_{22} = a_{11} \pm a_{12} + a_{13} \pm 2a_{16} + a_{19} \pm a_{20} + a_{21} \qquad (8.24d)$$

$$b_{23} = b_{32} = a_{14} \pm a_{15} \pm a_{17} + a_{18} \qquad (8.24e)$$

$$b_{33} = a_{22} \pm 2a_{23} + a_{24} \qquad (8.24f)$$

with the upper sign of the symbols $\pm$ referring to subspace S_1 and the lower sign to subspace S_4.

Subspaces S_2 and S_3

$$\mathbf{B} = [b_{11}] \qquad (8.25)$$

where

$$b_{11} = a_{11} \mp a_{12} - a_{13} \pm 2a_{16} - a_{19} \mp a_{20} + a_{21} \qquad (8.26)$$

with the upper sign of $\pm$ or $\mp$ referring to subspace S_2 and the lower sign to subspace S_3.

Subspaces S_{51} and S_{52}

$$\mathbf{B} = \begin{bmatrix} b_{11} & b_{12} & b_{13} & b_{14} \\ b_{21} & b_{22} & b_{23} & b_{24} \\ b_{31} & b_{32} & b_{33} & b_{34} \\ b_{41} & b_{42} & b_{43} & b_{44} \end{bmatrix} \qquad (8.27)$$

where, for both subspaces,

$$b_{11} = a_1 - a_{10} \qquad (8.28a)$$

$$b_{12} = b_{21} = a_5 - a_8 \qquad (8.28b)$$

$$b_{13} = b_{31} = 2a_2 - 2a_9 \qquad (8.28c)$$

$$b_{14} = b_{41} = 2a_3 - 2a_7 \tag{8.28d}$$

$$b_{22} = a_{22} - a_{24} \tag{8.28e}$$

$$b_{23} = b_{32} = 2a_{14} - 2a_{18} \tag{8.28f}$$

$$b_{24} = b_{42} = 2a_{15} - 2a_{17} \tag{8.28g}$$

$$b_{33} = a_{11} + a_{13} - a_{19} - a_{21} \tag{8.28h}$$

$$b_{34} = b_{43} = a_{12} - a_{20} \tag{8.28i}$$

$$b_{44} = a_{11} - a_{13} + a_{19} - a_{21} \tag{8.28j}$$

8.3.3 Eigenvalues

The diagonal mass matrices for the various subspaces, consisting of nonzero elements m_{ii} $(i = 1,2,...,r)$ as defined in Section 8.2.3, are written down from Figure 8.3(b), read in conjunction with Figure 8.4, as follows:

Subspaces S_1 and S_4

$$\mathbf{M} = \begin{bmatrix} m_1 & 0 & 0 \\ 0 & m_2 & 0 \\ 0 & 0 & m_3 \end{bmatrix} \tag{8.29}$$

Subspaces S_2 and S_3

$$\mathbf{M} = [m_2] \tag{8.30}$$

Subspaces S_{51} and S_{52}

$$\mathbf{M} = \begin{bmatrix} m_1 & 0 & 0 & 0 \\ 0 & m_3 & 0 & 0 \\ 0 & 0 & m_2 & 0 \\ 0 & 0 & 0 & m_2 \end{bmatrix} \tag{8.31}$$

Writing out the vanishing condition for eigenvalue determination – Equation (8.12) – for the various subspaces, we obtain

Subspaces S_1 and S_4

$$
\begin{vmatrix}
b_{11}-(\lambda/m_1) & b_{12} & b_{13} \\
b_{21} & b_{22}-(\lambda/m_2) & b_{23} \\
b_{31} & b_{32} & b_{33}-(\lambda/m_3)
\end{vmatrix} = 0 \tag{8.32}
$$

using the values of b_{ij} ($i = 1,2,3$; $j = 1,2,3$) appropriate for each subspace, as defined by Equations (8.24).

Subspaces S_2 and S_3

$$
b_{11} - (\lambda/m_2) = 0 \quad \Rightarrow \quad \lambda = b_{11}m_2 \tag{8.33}
$$

using the value of b_{11} appropriate for each subspace, as defined by Equation (8.26).

Subspace S_{51} and S_{52}

$$
\begin{vmatrix}
b_{11}-(\lambda/m_1) & b_{12} & b_{13} & b_{14} \\
b_{21} & b_{22}-(\lambda/m_3) & b_{23} & b_{24} \\
b_{31} & b_{32} & b_{33}-(\lambda/m_2) & b_{34} \\
b_{41} & b_{42} & b_{43} & b_{44}-(\lambda/m_2)
\end{vmatrix} = 0 \tag{8.34}
$$

where, for these two subspaces, the parameters b_{ij} ($i = 1,2,3,4$; $j = 1, 2,3,4$) are identical, being given by Equations (8.28) for either subspace. This implies that *subspace S_5 is always associated with doubly repeating roots for λ* (that is, four distinct roots in the present example), *so that only one of the subspaces S_{51} and S_{52} needs be considered in order to generate all the roots of subspace S_5*. This point will be elaborated further in Chapter 9.

Thus, compared with conventional analysis that requires the solution of a 16th-degree polynomial equation in λ in order to obtain all the eigenvalues for the 16-noded grid, use of group theory permits the required 16 eigenvalues of the problem to be determined much more easily through the solution of two third-degree polynomial equations, two very simple first-degree equations, and only one fourth-degree polynomial equation (generating four doubly repeating roots, that is, a total of eight roots), all independently of each other.

8.4 CONCLUSION

By decomposing a problem with n degrees of freedom into a number of independent subspaces spanned by symmetry-adapted functions, group theory enables the natural frequencies of vibration of symmetric configurations of grid-mass systems, typically those belonging to the symmetry groups C_{2v} and C_{4v}, to be obtained through the solution of a small number of mutually independent characteristic equations each of degree a fraction of n, with considerable savings in computational effort.

Chapter 9

High-tension cable nets

9.1 BASIC ASSUMPTIONS AND GEOMETRIC FORMULATION

Cable nets often consist of two families of pretensioned elastic cables intersecting each other at right angles when viewed in plan, with each cable being curved in a vertical plane [1, 2]. Although cable nets are generally considered to be geometrically nonlinear structures and usually analysed as such [3–6], their load-deformation response under certain conditions may be approximately linear (permitting a linear analysis), or if not, essentially linear techniques may be used to overcome the nonlinear effects [1, 7, 8].

It is assumed in the present treatment that the net is constructed with cables exclusively in tension, each cable lying in a vertical plane. The horizontal projection (on the xy plane) of the net comprises two sets of straight lines perpendicular to each other. The cables are connected to each other, and to rigid supports, through ideal, frictionless hinges. The cable net will be analysed as a discrete system, with the external loads and inertial forces concentrated at the joints. Only vertical loads (parallel to the z axis) are considered to act at the cable joints.

The horizontal component of the force in any given cable is assumed to be constant over the entire length of the cable. It is also assumed that the cables are highly tensioned, and that the cable-net system is relatively shallow and experiences vertical vibrations of small amplitude. Under these conditions, the horizontal component of the force in a given cable, invariant with respect to position along the cable, also remains practically constant with respect to time during vibration, permitting a linear analysis of the cable-net response.

It should be pointed out that the above modelling of cable-net behaviour has been made deliberately simple, even within the context of linear theory itself (in which more accurate physical models of cable-net behaviour already exist), in order to illustrate more clearly the proposed group-theoretic computational scheme. The techniques can be extended to more general models of cable nets, such as those where horizontal inertial forces are also taken into account, within the framework of linear theory.

For simplicity, we will consider only examples for which (1) the projections of the support points of the net on the horizontal plane all lie on a closed rectangle (or a closed square) and (2) the projections of the cables on the horizontal plane form lines that are equally spaced in both the x and the y directions. However, these restrictions on anchorage positions and cable spacings are not necessary for the proposed method to apply, as long as the overall configuration of the cable net exhibits symmetry properties.

In an xyz Cartesian coordinate reference system, denoting the constant horizontal component of the cable force in the kth cable of the x-oriented cables (numbering ξ) by H_{xk}, the constant horizontal component of the cable force in the jth cable of the y-oriented cables (numbering η) by H_{yj}, the horizontal spacing of the x-oriented cables by a, the horizontal spacing of the y-oriented cables by b, the vertical coordinate of the equilibrium position of joint (j,k) of the net by $z_{j,k}$, and the vertical load at joint (j,k) by $P_{j,k}$, the following relation, which is a statement of the condition of vertical equilibrium at each free joint (j,k) of the net ($j = 1, 2,\ldots,\eta$; $k = 1, 2,\ldots,\xi$), holds [9]:

$$\frac{H_{xk}}{a}(2z_{j,k} - z_{j-1,k} - z_{j+1,k}) + \frac{H_{yj}}{b}(2z_{j,k} - z_{j,k-1} - z_{j,k+1}) = P_{j,k} \qquad (9.1)$$

Now, for small-amplitude vertical vibrations of the cable net joints about their equilibrium positions, which will be denoted by the displacement $w_{j,k}$ (measured from the equilibrium position of joint (j,k)), the intervals a and b do not significantly change during vibration, and, as already stated, H_{xk} and H_{yj} also remain practically constant. Adapting expression (9.1), the equation of motion for joint (j,k) may be written as

$$\frac{H_{xk}}{a}\left[2(z_{j,k} + w_{j,k}) - (z_{j-1,k} + w_{j-1,k}) - (z_{j+1,k} + w_{j+1,k})\right]$$
$$+ \frac{H_{yj}}{b}\left[2(z_{j,k} + w_{j,k}) - (z_{j,k-1} + w_{j,k-1}) - (z_{j,k+1} + w_{j,k+1})\right] - P_{j,k} + m_{j,k}\ddot{w}_{j,k} = 0$$

which reduces to

$$\frac{H_{xk}}{a}(2w_{j,k} - w_{j-1,k} - w_{j+1,k}) + \frac{H_{yj}}{b}(2w_{j,k} - w_{j,k-1} - w_{j,k+1}) + m_{j,k}\ddot{w}_{j,k} = 0 \quad (9.2)$$

where $m_{j,k}$ is the concentrated mass at joint (j,k), and $\ddot{w}_{j,k}$ its acceleration with respect to time.

Adopting a conventional flexibility formulation of the problem, the associated flexibility coefficients can readily be calculated, on the basis of Equation (9.1), as the deflections that ensue at each of the $\eta\xi$ inner joints of the net, when a unit vertical load is applied at joint (r,s) with all other joints unloaded. We set up, using Equation (9.1), a system of $\eta\xi$ simultaneous equations in $\eta\xi$ unknowns (that is, the $z_{j,k}$'s), whose right-hand sides are all zeros except when $j = r$ and $k = s$, when $P_{j,k} = P_{r,s} = 1$; this system is then solved for the $z_{j,k}$'s.

The procedure is repeated with the unit vertical load at another joint of the net, until all joints of the net have been subjected, one joint at a time, to the unit vertical load. In this way, all the $\eta^2\,\xi^2$ flexibility coefficients of the linear-elastic net can be generated. If η and ξ are both large, the procedure can be computationally expensive.

On the other hand, the group-theoretic approach decomposes the vector space of the structural problem into a number of independent subspaces spanned by symmetry-adapted variables, within which flexibility matrices are of dimensions much smaller than $\eta\xi \times \eta\xi$, and characteristic polynomial equations are of degree a fraction of $\eta\xi$. This permits eigenvalues, eigenvectors and mode shapes for the entire problem to be obtained much more efficiently than in the case of the conventional approach.

The fundamental requirement for the group-theoretic computational scheme to be applicable is that the overall configuration of the cable net (that is, arrangement of cables, pattern of cable prestress forces, pattern of support types) must possess one or more symmetry properties.

9.2 OUTLINE OF COMPUTATIONAL SCHEME

For any system with one or more symmetry properties, the first step of a group-theoretic analysis consists in identifying the symmetry elements, and hence the symmetry group of the configuration.

In vibration problems, and as already seen in earlier chapters, applying an idempotent to each of the arbitrary functions $\{\phi_1, \phi_2, ..., \phi_n\}$ describing the motions of a system with n degrees of freedom (whose positions are permuted by the symmetry elements of the group) gives the symmetry-adapted functions for the corresponding subspace, from which a set of r ($r \ll n$) independent basis vectors spanning that subspace may readily be selected. All the eigenvalues of the problem may then be obtained by considering each of such subspaces separately, and solving the characteristic polynomial equation associated with the subspace independently of those associated with the other subspaces.

Since the degree of any one of these subspace polynomial equations (which equals the dimension r of the associated subspace) is only a fraction of n, it is evident that the group-theoretic procedure always affords substantial reductions in computational effort, a benefit that was amply illustrated in Chapters 7 and 8.

In general, if the n-dimensional vector space of a particular problem decomposes into k independent subspaces of dimensions $\{r_1, r_2, ..., r_k\}$, where $r_1 + r_2 + \ldots + r_k = n$, then an approximate measure of the computational efficiency of the solution stage (expressed as a ratio of the computational effort required to solve k polynomial equations of degrees $\{r_1, r_2, \ldots, r_k\}$ to the computational effort required to solve one polynomial equation of degree n) is given by $\{r_1^2 + r_2^2 + \ldots + r_k^2\}/n^2$; the precise measure depends, of course, on the numerical technique employed for the solution of polynomials.

Taking each of the now known eigenvalues in turn, the eigenvector associated with it is obtained by considering only the subspace from which that eigenvalue was calculated. Substituting the eigenvalue into the eigenvalue equation for the parent r-dimensional subspace, we solve for r eigenvector components corresponding to a particular natural frequency of the cable net, which gives the *eigenvector in the r-dimensional subspace*.

For this frequency, the *eigenvector in the original n-dimensional vector space of the problem* is simply obtained by allocating the calculated value of each subspace-eigenvector component to all the cable-net joints associated with the basis vector corresponding to the subspace-eigenvector component, with the signs (positive or negative) of the allocations being in accordance with those of the basis-vector coordinates.

In decomposing the problem, the number of independent subproblems to be expected is, of course, given by the number of classes of the applicable symmetry group, as already explained in Chapter 6. Thus, in analysing a single cable with one plane of symmetry (at midspan), for which the group C_{1v} is applicable, only two subproblems, corresponding to the symmetric and the antisymmetric subspaces, are possible. In the case of a cable net rectangular in plan (group C_{2v}), four independent subproblems result, while for one with a square plan (group C_{4v}), the decomposition results in five independent subproblems.

Where more than one symmetry group is applicable, we must choose the symmetry group with the largest number of elements since, in general, this would ensure the maximum possible decomposition (that is, the greatest number of subproblems of the smallest possible dimensions). Cable nets can have a wide variety of configurations corresponding to different symmetry groups, but the number of subproblems into which the original problem can be decomposed is always limited by the number of classes associated with its group, which ranges from 2 to 12 for all problems likely to be encountered in practice.

9.3 ILLUSTRATIVE EXAMPLES

Let us consider two examples of nonplanar cable nets formed by two families of high-tension cables intersecting perpendicularly on the xy horizontal projection. These are depicted in plan in Figure 9.1, in which diagram (a) shows a 24-node configuration belonging to the symmetry group C_{2v},

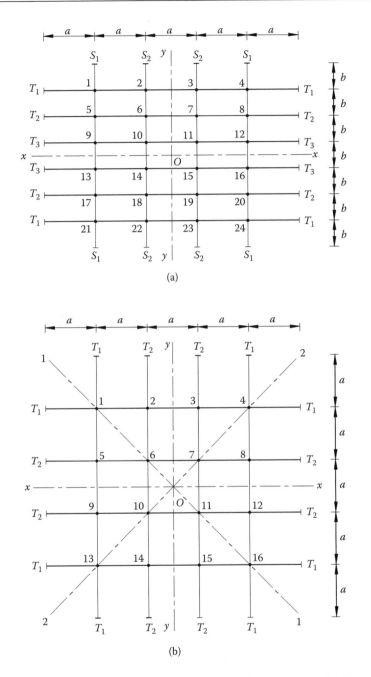

Figure 9.1 Horizontal projections of cable nets, showing adopted node numbering, vertical planes of symmetry and pattern of prestressing: (a) 24-node C_{2v} configuration, (b) 16-node C_{4v} configuration.

while diagram (b) shows a 16-node configuration belonging to the symmetry group C_{4v}. For each configuration, the centre of symmetry is denoted by O. Also shown in the diagrams are the adopted numbering of nodes and labelling of planes of reflection.

For the first example, the horizontal projections of the cable ends lie on a closed rectangle, while for the second example, the horizontal projections of the cable ends lie on a closed square. The horizontal spacing between cables is constant for cables in a given direction, being as shown in the figure. The horizontal projection of the cable prestress force is constant for a given cable, but allowed to differ from one cable to another while still preserving the full symmetry of the structural configuration.

9.4 SYMMETRY-ADAPTED FUNCTIONS

9.4.1 Rectangular 24-node configuration

As earlier indicated, the symmetry-adapted positional functions for a given subspace S_l ($l = 1, 2, 3, 4$) are obtained by first applying the corresponding idempotent P_l (given by one of Equations (6.3)) on each positional function ϕ_i ($i = 1, 2,..., 24$) of the problem, and noting the results. Selecting any full set of linearly independent symmetry-adapted positional functions then gives the set of basis vectors spanning that subspace.

Consider subspace S_1 first. As a reminder of how the results are obtained, the term $C_2\phi_1$ (one of the four terms of the expansion for $P_1\phi_1$) means 'apply a C_2 operation (that is, a 180° rotation about O) on position 1' (refer to Figure 9.1(a)), which gives the result ϕ_{24} (that is, position 1 is moved to position 24 by the operation C_2). Applying P_1 to each of the 24 positions of the cable net, we obtain:

Subspace S_1

$$P_1\phi_1 = \frac{1}{4}(e + C_2 + \sigma_x + \sigma_y)\phi_1$$

$$= \frac{1}{4}(\phi_1 + \phi_{24} + \phi_{21} + \phi_4) = P_1\phi_4 = P_1\phi_{21} = P_1\phi_{24}$$

$$P_1\phi_2 = \frac{1}{4}(e + C_2 + \sigma_x + \sigma_y)\phi_2$$

$$= \frac{1}{4}(\phi_2 + \phi_{23} + \phi_{22} + \phi_3) = P_1\phi_3 = P_1\phi_{22} = P_1\phi_{23}$$

$$P_1\phi_5 = \frac{1}{4}(e + C_2 + \sigma_x + \sigma_y)\phi_5$$

$$= \frac{1}{4}(\phi_5 + \phi_{20} + \phi_{17} + \phi_8) = P_1\phi_8 = P_1\phi_{17} = P_1\phi_{20}$$

$$P_1\phi_6 = \frac{1}{4}(e + C_2 + \sigma_x + \sigma_y)\phi_6$$

$$= \frac{1}{4}(\phi_6 + \phi_{19} + \phi_{18} + \phi_7) = P_1\phi_7 = P_1\phi_{18} = P_1\phi_{19}$$

$$P_1\phi_9 = \frac{1}{4}(e + C_2 + \sigma_x + \sigma_y)\phi_9$$

$$= \frac{1}{4}(\phi_9 + \phi_{16} + \phi_{13} + \phi_{12}) = P_1\phi_{12} = P_1\phi_{13} = P_1\phi_{16}$$

$$P_1\phi_{10} = \frac{1}{4}(e + C_2 + \sigma_x + \sigma_y)\phi_{10}$$

$$= \frac{1}{4}(\phi_{10} + \phi_{15} + \phi_{14} + \phi_{11}) = P_1\phi_{11} = P_1\phi_{14} = P_1\phi_{15}$$

Thus, S_1 is six-dimensional (there are only six linearly independent results). The basis vectors for the subspace may be taken as follows:

$$\Phi_1 = \phi_1 + \phi_4 + \phi_{21} + \phi_{24} \tag{9.3a}$$

$$\Phi_2 = \phi_2 + \phi_3 + \phi_{22} + \phi_{23} \tag{9.3b}$$

$$\Phi_3 = \phi_5 + \phi_8 + \phi_{17} + \phi_{20} \tag{9.3c}$$

$$\Phi_4 = \phi_6 + \phi_7 + \phi_{18} + \phi_{19} \tag{9.3d}$$

$$\Phi_5 = \phi_9 + \phi_{12} + \phi_{13} + \phi_{16} \tag{9.3e}$$

$$\Phi_6 = \phi_{10} + \phi_{11} + \phi_{14} + \phi_{15} \tag{9.3f}$$

The basis vectors for the remaining three subspaces are obtained in a similar manner, using the idempotent P_2 in the case of subspace S_2, P_3 in the case of subspace S_3, and P_4 in the case of subspace S_4. Each of these subspaces is also found to be six-dimensional; the independent basis vectors may be taken as follows:

Subspace S_2

$$\Phi_1 = \phi_1 - \phi_4 - \phi_{21} + \phi_{24} \tag{9.4a}$$

$$\Phi_2 = \phi_2 - \phi_3 - \phi_{22} + \phi_{23} \tag{9.4b}$$

$$\Phi_3 = \phi_5 - \phi_8 - \phi_{17} + \phi_{20} \tag{9.4c}$$

$$\Phi_4 = \phi_6 - \phi_7 - \phi_{18} + \phi_{19} \tag{9.4d}$$

$$\Phi_5 = \phi_9 - \phi_{12} - \phi_{13} + \phi_{16} \tag{9.4e}$$

$$\Phi_6 = \phi_{10} - \phi_{11} - \phi_{14} + \phi_{15} \tag{9.4f}$$

Subspace S_3

$$\Phi_1 = \phi_1 - \phi_4 + \phi_{21} - \phi_{24} \tag{9.5a}$$

$$\Phi_2 = \phi_2 - \phi_3 + \phi_{22} - \phi_{23} \tag{9.5b}$$

$$\Phi_3 = \phi_5 - \phi_8 + \phi_{17} - \phi_{20} \tag{9.5c}$$

$$\Phi_4 = \phi_6 - \phi_7 + \phi_{18} - \phi_{19} \tag{9.5d}$$

$$\Phi_5 = \phi_9 - \phi_{12} + \phi_{13} - \phi_{16} \tag{9.5e}$$

$$\Phi_6 = \phi_{10} - \phi_{11} + \phi_{14} - \phi_{15} \tag{9.5f}$$

Subspace S_4

$$\Phi_1 = \phi_1 + \phi_4 - \phi_{21} - \phi_{24} \tag{9.6a}$$

$$\Phi_2 = \phi_2 + \phi_3 - \phi_{22} - \phi_{23} \tag{9.6b}$$

$$\Phi_3 = \phi_5 + \phi_8 - \phi_{17} - \phi_{20} \tag{9.6c}$$

$$\Phi_4 = \phi_6 + \phi_7 - \phi_{18} - \phi_{19} \tag{9.6d}$$

$$\Phi_5 = \phi_9 + \phi_{12} - \phi_{13} - \phi_{16} \tag{9.6e}$$

$$\Phi_6 = \phi_{10} + \phi_{11} - \phi_{14} - \phi_{15} \tag{9.6f}$$

All basis vectors of a given subspace have the same symmetry type. We may see this more clearly by plotting the four coordinates of each basis vector (as unit positive or negative quantities, depending on the sign of the coordinate) onto the corresponding node in Figure 9.1(a), and then noting the orientation of the symmetry and antisymmetry planes of the ensuing pattern. Symmetry plots of basis vectors will be illustrated in due course, in connection with the C_{4v} example.

9.4.2 Square 16-node configuration

Applying the idempotent P_l (l = 1, 2, 3, 4, 5 – see Equations (6.4)) as a positional operator on each of the 16 positional functions $\{\phi_1, \phi_2, ..., \phi_{16}\}$ of the second example of Figure 9.1, we obtain symmetry-adapted positional functions for subspace S_l, from which the basis vectors for the subspace may be written down. It is found that subspaces S_1 and S_4 are each three-dimensional, subspaces S_2 and S_3 are each one-dimensional, while subspace S_5 is eight-dimensional, as follows:

Subspace S_1

$$\Phi_1 = \phi_1 + \phi_4 + \phi_{13} + \phi_{16} \tag{9.7a}$$

$$\Phi_2 = \phi_2 + \phi_3 + \phi_5 + \phi_8 + \phi_9 + \phi_{12} + \phi_{14} + \phi_{15} \tag{9.7b}$$

$$\Phi_3 = \phi_6 + \phi_7 + \phi_{10} + \phi_{11} \tag{9.7c}$$

Subspace S_2

$$\Phi_1 = \phi_2 - \phi_3 - \phi_5 + \phi_8 + \phi_9 - \phi_{12} - \phi_{14} + \phi_{15} \tag{9.8}$$

Subspace S_3

$$\Phi_1 = \phi_2 + \phi_3 - \phi_5 - \phi_8 - \phi_9 - \phi_{12} + \phi_{14} + \phi_{15} \tag{9.9}$$

Subspace S_4

$$\Phi_1 = \phi_1 - \phi_4 - \phi_{13} + \phi_{16} \tag{9.10a}$$

$$\Phi_2 = \phi_2 - \phi_3 + \phi_5 - \phi_8 - \phi_9 + \phi_{12} - \phi_{14} + \phi_{15} \tag{9.10b}$$

$$\Phi_3 = \phi_6 - \phi_7 - \phi_{10} + \phi_{11} \tag{9.10c}$$

Subspace S_5

$$\Phi_1 = \phi_1 - \phi_{16} \tag{9.11a}$$

$$\Phi_2 = \phi_2 - \phi_{15} \tag{9.11b}$$

$$\Phi_3 = \phi_3 - \phi_{14} \tag{9.11c}$$

$$\Phi_4 = \phi_4 - \phi_{13} \tag{9.11d}$$

$$\Phi_5 = \phi_5 - \phi_{12} \tag{9.11e}$$

$$\Phi_6 = \phi_6 - \phi_{11} \tag{9.11f}$$

$$\Phi_7 = \phi_7 - \phi_{10} \tag{9.11g}$$

$$\Phi_8 = \phi_8 - \phi_9 \tag{9.11h}$$

Subspace S_5 can be further decomposed into two independent four-dimensional subspaces S_{51} and S_{52}, spanned by new basis vectors obtained by linearly combining the vectors of Equations (9.11) in such a way as to form two orthogonal sets:

Subspace S_{51}

$$\Phi_1' = \Phi_1 = \phi_1 - \phi_{16} \tag{9.12a}$$

$$\Phi_2' = \Phi_6 = \phi_6 - \phi_{11} \tag{9.12b}$$

$$\Phi_3' = \Phi_3 - \Phi_8 = \phi_3 - \phi_8 + \phi_9 - \phi_{14} \tag{9.12c}$$

$$\Phi_4' = \Phi_2 + \Phi_5 = \phi_2 + \phi_5 - \phi_{12} - \phi_{15} \tag{9.12d}$$

Subspace S_{52}

$$\Phi_1' = \Phi_4 = \phi_4 - \phi_{13} \tag{9.13a}$$

$$\Phi_2' = \Phi_7 = \phi_7 - \phi_{10} \tag{9.13b}$$

$$\Phi_3' = \Phi_3 + \Phi_8 = \phi_3 + \phi_8 - \phi_9 - \phi_{14} \tag{9.13c}$$

$$\Phi_4' = \Phi_2 - \Phi_5 = \phi_2 - \phi_5 + \phi_{12} - \phi_{15} \tag{9.13d}$$

The orthogonality of the two vector sets (9.12) and (9.13) may be seen by glancing ahead at pages 182 and 183 (figure to be introduced shortly), diagrams (e) and (f), in which solid and dashed lines have been marked at the centre of the sketches, to indicate the orientation of symmetry and antisymmetry planes, respectively, of the basis-vector plots. Orientations of symmetry and antisymmetry planes for subspace S_{51} are seen to be perpendicular to their counterparts for subspace S_{52}, illustrating the orthogonality of the two sets of vectors. The same plots also illustrate the general property that all basis vectors of a given subspace have the same symmetry type.

In general, the above decomposition of subspace S_5 into subspaces S_{51} and S_{52}, and the derivation of the associated basis vectors for a given problem, may be achieved more systematically by splitting idempotent P_5 (Equation 6.4e) into components P_{51} and P_{52}, such that $P_{51} + P_{52} = P_5$, $P_{51} P_{51} = P_{51}$, $P_{52} P_{52} = P_{52}$ and $P_{51} P_{52} = 0$. These conditions are satisfied by the operators

$$P_{51} = \frac{1}{4}(e - C_2 + \sigma_1 - \sigma_2) \tag{9.14a}$$

$$P_{52} = \frac{1}{4}(e - C_2 - \sigma_1 + \sigma_2) \tag{9.14b}$$

which may also be seen to be each orthogonal to the idempotents P_i $(i = 1, 2, 3, 4)$ given by expressions (6.4a–d). Thus, in the present example, applying P_{51} and P_{52} in turn as positional operators on each positional function ϕ_j ($j =1, 2, ..., 16$) generates the basis vectors for subspace S_{51} (as given by expressions (9.12)) and subspace S_{52} (as given by expressions (9.13)) automatically.

It must be noted that the indicated decomposition of subspace S_5 into subspaces S_{51} and S_{52}, by means of the operators in Equation (9.14), is applicable for all eigenvalue problems. A physical explanation of this is given in the next section.

Strictly speaking, subspaces S_{51} and S_{52} are not group-invariant subspaces in the sense explained in Chapter 6, because (1) they originate from *the same irreducible representation R_5*, and yet to each group-invariant subspace S_i of a vector space V must correspond a distinct irreducible representation R_i, and (2) their *symmetry types are physically indistinguishable*, and yet it must be possible to distinguish each group-invariant subspace of a physical system by the symmetry type of the basis vectors spanning it. The second point will become clearer in the next section.

9.5 SYMMETRY-ADAPTED FLEXIBILITY MATRICES

9.5.1 Basis-vector plots and subspace properties

Figure 9.2 (referring to the 24-node C_{2v} configuration) and Figure 9.3 (referring to the 16-node C_{4v} configuration) show, for each subspace of the associated problem, unit vertical forces applied upon the cable-net nodes in accordance with the coordinates of the respective basis vectors. The notation $\otimes$ and $\odot$ in these diagrams denotes downward (that is, positive) and upward (that is, negative) unit forces, respectively.

For clarity, the sets of unit vertical forces associated with the various basis vectors of a given subspace are shown on separate diagrams. Also, to avoid cluttering the diagrams, the numbering of the nodes is not shown; however,

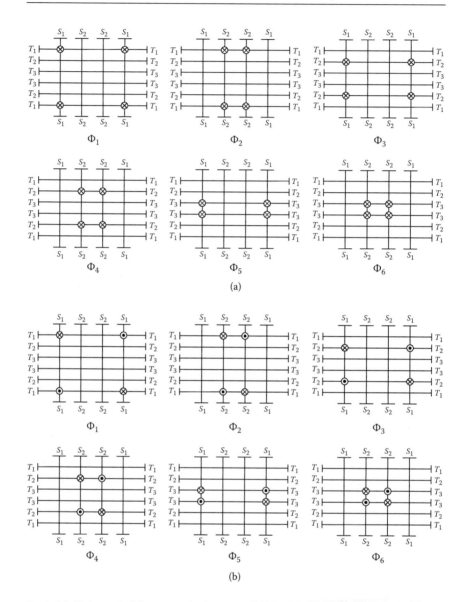

Figure 9.2 Unit vertical forces applied in accordance with the coordinates of the basis vectors, for the 24-node C_{2v} configuration: (a) subspace S_1, (b) subspace S_2.

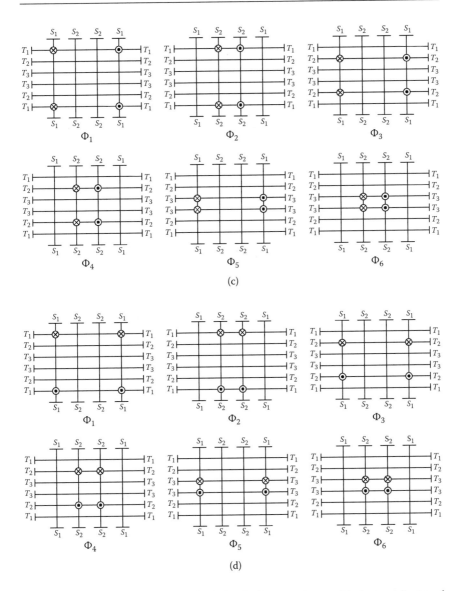

Figure 9.2 (Continued) Unit vertical forces applied in accordance with the coordinates of the basis vectors, for the 24-node C_{2v} configuration: (c) subspace S_3, (d) subspace S_4.

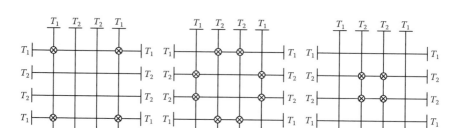

(a)

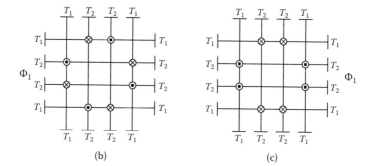

(b)   (c)

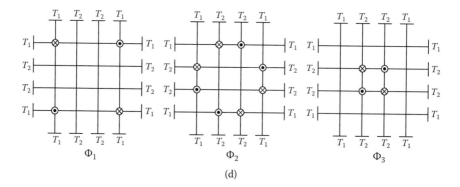

(d)

Figure 9.3 Unit vertical forces applied in accordance with the coordinates of the basis vectors, for the 16-node C_{4v} configuration: (a) subspace S_1, (b) subspace S_2, (c) subspace S_3, (d) subspace S_4.

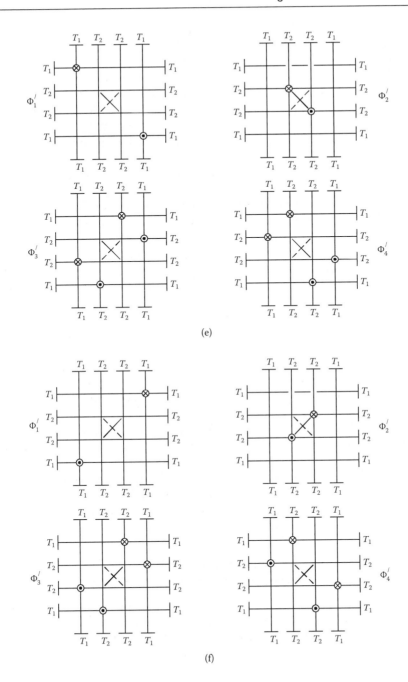

Figure 9.3 (Continued) Unit vertical forces applied in accordance with the coordinates of the basis vectors, for the 16-node C_{4v} configuration: (e) subspace S_{51}, (f) subspace S_{52}.

these diagrams have been drawn with exactly the same orientation as in Figure 9.1, to which reference may be made as regards the identities of the point-load locations depicted in Figures 9.2 and 9.3.

When all the r basis-vector plots for a given subspace, shown separately in Figures 9.2 and 9.3, are superimposed on the same diagram, with all cable nodes associated with a basis vector Φ_i now requiring to be labelled i for identification purposes, any nodes that remain unlabelled (that is, nodes that are not associated with any one of the r basis vectors of the subspace) imply *stationary* points. Such nodes are seen to lie in *antisymmetry* planes of the pattern of the superimposed basis-vector plots.

No instance of stationary nodes occurs in the case of the 24-node C_{2v} example. In the case of the 16-node C_{4v} example, on the other hand, subspaces S_2 and S_3 are each associated with eight stationary nodes (the points lying on *either* of the principal diagonals of the square net), while subspaces S_{51} and S_{52} are each associated with the occurrence of four stationary nodes (the points on *one* of the two principal diagonals of the square net).

The set of basis vectors for subspace S_{51} is physically indistinguishable from that for subspace S_{52}: for every vector in S_{51}, there is one exactly identical to it in S_{52}, except for orientation (compare Figures 9.3(e, f)). As the physical properties of a physical system, such as natural frequencies and modes of deformation, are not affected by its orientation in space, a solution for any required physical properties based on subspace S_{51} must be identical to that based on subspace S_{52}.

Thus, as has also been noted in Section 8.3, only one of the two subspaces S_{51} and S_{52} actually needs to be considered in any vibration problem involving the C_{4v} symmetry group, as these S_5 subspaces are always associated with the same set of eigenvalues (which, in the conventional sense, constitute doubly repeating roots of the problem). In further pursuing our example of the 16-node cable net, we will take subspace S_{51} as representative of the two identical subspaces of S_5.

9.5.2 Equilibrium considerations and symmetry-adapted flexibility coefficients

For a given subspace spanned by r basis vectors (for example, $r = 6$ for each of the four subspaces of the 24-node C_{2v} example), let a_{ij} ($i = 1, 2,...,r$; $j = 1, 2,...,r$) be the vertical-displacement magnitude at *any* of the nodes of the basis vector Φ_i, due to unit vertical forces applied at *all* the nodes of the basis vector Φ_j.

The condition of vertical force equilibrium at each of the r sets of nodes (corresponding to the r basis vectors of the subspace) – Equation (9.1) gives this condition for an arbitrary node – leads to r simultaneous equations in the r static deflection unknowns $\{a_{1j}, a_{2j}, \ldots, a_{rj}\}$, which may be expressed as follows:

$$
\begin{bmatrix}
b_{11} & b_{12} & . & b_{1r} \\
b_{21} & b_{22} & . & b_{2r} \\
. & . & . & . \\
b_{r1} & b_{r2} & . & b_{rr}
\end{bmatrix}
\begin{bmatrix}
a_{1j} \\
a_{2j} \\
. \\
a_{rj}
\end{bmatrix}
=
\begin{bmatrix}
\delta_{1j} \\
\delta_{2j} \\
. \\
\delta_{rj}
\end{bmatrix}
\tag{9.15}
$$

for $j = 1, 2, \ldots, r$. For the right-hand side, $\delta_{ij} = 1$ if $i = j$, and $\delta_{ij} = 0$ if $i \neq j$. This equation can be written as

$$
\mathbf{B}_l \, \mathbf{A}_j = \boldsymbol{\delta}_j
\tag{9.16}
$$

where $\mathbf{B}_l$ is an $r \times r$ matrix corresponding to subspace S_l (we will refer to this as the *static equilibrium matrix* for the subspace), $\boldsymbol{\delta}_j$ is an $r \times 1$ column vector consisting of a 1 at row j and zeros everywhere else, and $\mathbf{A}_j$ is the $r \times 1$ column vector of the deflections corresponding to the application of unit vertical forces at each of the nodes of Φ_j. These deflections are, by definition, the flexibility coefficients for the subspace.

Rearranging Equation (9.16) then yields the column vector $\mathbf{A}_j$ of the flexibility matrix $\mathbf{A}_l$ for subspace S_l. When the $\mathbf{A}_j$ ($j = 1, 2, \ldots, r$) are put together, they form the full flexibility matrix $\mathbf{A}_l$ for subspace S_l:

$$
\mathbf{A}_j = [\mathbf{B}_l]^{-1} \, \boldsymbol{\delta}_j; \, j = 1, 2, \ldots, r
\tag{9.17}
$$

$$
\mathbf{A}_l = [\mathbf{A}_1 \mid \mathbf{A}_2 \mid \ldots \mid \mathbf{A}_r]
\tag{9.18}
$$

Thus, once the static equilibrium matrices $\mathbf{B}_l$ are known, the subspace flexibility matrices follow from Equations (9.17) and (9.18). For the various subspaces of the two cable nets, the $\mathbf{B}_l$ matrices are derived below, by first expressing, on the basis of Equation (9.1), the individual conditions of vertical equilibrium at the nodes of each basis vector of a subspace, and then collecting these into the form $\mathbf{B}_l \, \mathbf{A}_j = \boldsymbol{\delta}_j$.

9.5.3 Derivation of the static equilibrium matrices

9.5.3.1 Rectangular 24-node configuration

9.5.3.1.1 Subspace S_l

With reference to Figure 9.2(a), application of Φ_j ($j = 1, 2, \ldots, 6$) yields the following nodal equations for force equilibrium in the vertical direction:

At nodes of Φ_1: $\dfrac{T_1}{a}(2a_{1j} - 0 - a_{2j}) + \dfrac{S_1}{b}(2a_{1j} - 0 - a_{3j}) = \delta_{1j}$

At nodes of Φ_2: $\dfrac{T_1}{a}(2a_{2j} - a_{1j} - a_{2j}) + \dfrac{S_2}{b}(2a_{2j} - 0 - a_{4j}) = \delta_{2j}$

At nodes of Φ_3: $\dfrac{T_2}{a}(2a_{3j} - 0 - a_{4j}) + \dfrac{S_1}{b}(2a_{3j} - a_{1j} - a_{5j}) = \delta_{3j}$

At nodes of Φ_4: $\dfrac{T_2}{a}(2a_{4j} - a_{3j} - a_{4j}) + \dfrac{S_2}{b}(2a_{4j} - a_{2j} - a_{6j}) = \delta_{4j}$

At nodes of Φ_5: $\dfrac{T_3}{a}(2a_{5j} - 0 - a_{6j}) + \dfrac{S_1}{b}(2a_{5j} - a_{3j} - a_{5j}) = \delta_{5j}$

At nodes of Φ_6: $\dfrac{T_3}{a}(2a_{6j} - a_{5j} - a_{6j}) + \dfrac{S_2}{b}(2a_{6j} - a_{4j} - a_{6j}) = \delta_{6j}$

that is,

$$
\begin{bmatrix}
2\left(\dfrac{T_1}{a}+\dfrac{S_1}{b}\right) & -\dfrac{T_1}{a} & -\dfrac{S_1}{b} & 0 & 0 & 0 \\[2mm]
-\dfrac{T_1}{a} & \dfrac{T_1}{a}+\dfrac{2S_2}{b} & 0 & -\dfrac{S_2}{b} & 0 & 0 \\[2mm]
-\dfrac{S_1}{b} & 0 & 2\left(\dfrac{T_2}{a}+\dfrac{S_1}{b}\right) & -\dfrac{T_2}{a} & -\dfrac{S_1}{b} & 0 \\[2mm]
0 & -\dfrac{S_2}{b} & -\dfrac{T_2}{a} & \dfrac{T_2}{a}+\dfrac{2S_2}{b} & 0 & -\dfrac{S_2}{b} \\[2mm]
0 & 0 & -\dfrac{S_1}{b} & 0 & \dfrac{2T_3}{a}+\dfrac{S_1}{b} & -\dfrac{T_3}{a} \\[2mm]
0 & 0 & 0 & -\dfrac{S_2}{b} & -\dfrac{T_3}{a} & \dfrac{T_3}{a}+\dfrac{S_2}{b}
\end{bmatrix}
\begin{bmatrix} a_{1j} \\ a_{2j} \\ a_{3j} \\ a_{4j} \\ a_{5j} \\ a_{6j} \end{bmatrix}
=
\begin{bmatrix} \delta_{1j} \\ \delta_{2j} \\ \delta_{3j} \\ \delta_{4j} \\ \delta_{5j} \\ \delta_{6j} \end{bmatrix}
$$

(9.19)

where, for $i = 1,2,...,6$, $\delta_{ij} = 1$ if $i = j$; $\delta_{ij} = 0$ if $i \neq j$.

9.5.3.1.2 Subspace S_2

With reference to Figure 9.2(b), application of Φ_j ($j = 1,2,...,6$) yields the following nodal equations for force equilibrium in the vertical direction:

At nodes of Φ_1: $\dfrac{T_1}{a}(2a_{1j} - 0 - a_{2j}) + \dfrac{S_1}{b}(2a_{1j} - 0 - a_{3j}) = \delta_{1j}$

At nodes of Φ_2: $\dfrac{T_1}{a}(2a_{2j} - a_{1j} + a_{2j}) + \dfrac{S_2}{b}(2a_{2j} - 0 - a_{4j}) = \delta_{2j}$

At nodes of Φ_3: $\dfrac{T_2}{a}(2a_{3j} - 0 - a_{4j}) + \dfrac{S_1}{b}(2a_{3j} - a_{1j} - a_{5j}) = \delta_{3j}$

At nodes of Φ_4: $\dfrac{T_2}{a}(2a_{4j} - a_{3j} + a_{4j}) + \dfrac{S_2}{b}(2a_{4j} - a_{2j} - a_{6j}) = \delta_{4j}$

At nodes of Φ_5: $\dfrac{T_3}{a}(2a_{5j} - 0 - a_{6j}) + \dfrac{S_1}{b}(2a_{5j} - a_{3j} + a_{5j}) = \delta_{5j}$

At nodes of Φ_6: $\dfrac{T_3}{a}(2a_{6j} - a_{5j} + a_{6j}) + \dfrac{S_2}{b}(2a_{6j} - a_{4j} + a_{6j}) = \delta_{6j}$

that is,

$$
\begin{bmatrix}
2\left(\dfrac{T_1}{a}+\dfrac{S_1}{b}\right) & -\dfrac{T_1}{a} & -\dfrac{S_1}{b} & 0 & 0 & 0 \\[2ex]
-\dfrac{T_1}{a} & \dfrac{3T_1}{a}+\dfrac{2S_2}{b} & 0 & -\dfrac{S_2}{b} & 0 & 0 \\[2ex]
-\dfrac{S_1}{b} & 0 & 2\left(\dfrac{T_2}{a}+\dfrac{S_1}{b}\right) & -\dfrac{T_2}{a} & -\dfrac{S_1}{b} & 0 \\[2ex]
0 & -\dfrac{S_2}{b} & -\dfrac{T_2}{a} & \dfrac{3T_2}{a}+\dfrac{2S_2}{b} & 0 & -\dfrac{S_2}{b} \\[2ex]
0 & 0 & -\dfrac{S_1}{b} & 0 & \dfrac{2T_3}{a}+\dfrac{3S_1}{b} & -\dfrac{T_3}{a} \\[2ex]
0 & 0 & 0 & -\dfrac{S_2}{b} & -\dfrac{T_3}{a} & 3\left(\dfrac{T_3}{a}+\dfrac{S_2}{b}\right)
\end{bmatrix}
\begin{bmatrix} a_{1j} \\ a_{2j} \\ a_{3j} \\ a_{4j} \\ a_{5j} \\ a_{6j} \end{bmatrix}
=
\begin{bmatrix} \delta_{1j} \\ \delta_{2j} \\ \delta_{3j} \\ \delta_{4j} \\ \delta_{5j} \\ \delta_{6j} \end{bmatrix}
$$

$$(9.20)$$

with δ_{ij} defined as before.

9.5.3.1.3 Subspace S_3

With reference to Figure 9.2(c), application of Φ_j ($j = 1,2,...,6$) yields the following nodal equations for force equilibrium in the vertical direction:

At nodes of Φ_1: $\dfrac{T_1}{a}(2a_{1j} - 0 - a_{2j}) + \dfrac{S_1}{b}(2a_{1j} - 0 - a_{3j}) = \delta_{1j}$

At nodes of Φ_2: $\dfrac{T_1}{a}(2a_{2j} - a_{1j} + a_{2j}) + \dfrac{S_2}{b}(2a_{2j} - 0 - a_{4j}) = \delta_{2j}$

At nodes of Φ_3: $\dfrac{T_2}{a}(2a_{3j} - 0 - a_{4j}) + \dfrac{S_1}{b}(2a_{3j} - a_{1j} - a_{5j}) = \delta_{3j}$

At nodes of Φ_4: $\dfrac{T_2}{a}(2a_{4j} - a_{3j} + a_{4j}) + \dfrac{S_2}{b}(2a_{4j} - a_{2j} - a_{6j}) = \delta_{4j}$

At nodes of Φ_5: $\dfrac{T_3}{a}(2a_{5j} - 0 - a_{6j}) + \dfrac{S_1}{b}(2a_{5j} - a_{3j} - a_{5j}) = \delta_{5j}$

At nodes of Φ_6: $\dfrac{T_3}{a}(2a_{6j} - a_{5j} + a_{6j}) + \dfrac{S_2}{b}(2a_{6j} - a_{4j} - a_{6j}) = \delta_{6j}$

that is,

$$
\begin{bmatrix}
2\left(\dfrac{T_1}{a} + \dfrac{S_1}{b}\right) & -\dfrac{T_1}{a} & -\dfrac{S_1}{b} & 0 & 0 & 0 \\[2ex]
-\dfrac{T_1}{a} & \dfrac{3T_1}{a} + \dfrac{2S_2}{b} & 0 & -\dfrac{S_2}{b} & 0 & 0 \\[2ex]
-\dfrac{S_1}{b} & 0 & 2\left(\dfrac{T_2}{a} + \dfrac{S_1}{b}\right) & -\dfrac{T_2}{a} & -\dfrac{S_1}{b} & 0 \\[2ex]
0 & -\dfrac{S_2}{b} & -\dfrac{T_2}{a} & \dfrac{3T_2}{a} + \dfrac{2S_2}{b} & 0 & -\dfrac{S_2}{b} \\[2ex]
0 & 0 & -\dfrac{S_1}{b} & 0 & \dfrac{2T_3}{a} + \dfrac{S_1}{b} & -\dfrac{T_3}{a} \\[2ex]
0 & 0 & 0 & -\dfrac{S_2}{b} & -\dfrac{T_3}{a} & \dfrac{3T_3}{a} + \dfrac{S_2}{b}
\end{bmatrix}
\begin{bmatrix} a_{1j} \\ a_{2j} \\ a_{3j} \\ a_{4j} \\ a_{5j} \\ a_{6j} \end{bmatrix}
=
\begin{bmatrix} \delta_{1j} \\ \delta_{2j} \\ \delta_{3j} \\ \delta_{4j} \\ \delta_{5j} \\ \delta_{6j} \end{bmatrix}
$$

$$(9.21)$$

with δ_{ij} defined as before.

9.5.3.1.4 Subspace S_4

With reference to Figure 9.2(d), application of Φ_j ($j = 1, 2, \ldots, 6$) yields the following nodal equations for force equilibrium in the vertical direction:

At nodes of Φ_1: $\dfrac{T_1}{a}(2a_{1j} - 0 - a_{2j}) + \dfrac{S_1}{b}(2a_{1j} - 0 - a_{3j}) = \delta_{1j}$

At nodes of Φ_2: $\dfrac{T_1}{a}(2a_{2j} - a_{1j} - a_{2j}) + \dfrac{S_2}{b}(2a_{2j} - 0 - a_{4j}) = \delta_{2j}$

At nodes of Φ_3: $\dfrac{T_2}{a}(2a_{3j}-0-a_{4j})+\dfrac{S_1}{b}(2a_{3j}-a_{1j}-a_{5j})=\delta_{3j}$

At nodes of Φ_4: $\dfrac{T_2}{a}(2a_{4j}-a_{3j}-a_{4j})+\dfrac{S_2}{b}(2a_{4j}-a_{2j}-a_{6j})=\delta_{4j}$

At nodes of Φ_5: $\dfrac{T_3}{a}(2a_{5j}-0-a_{6j})+\dfrac{S_1}{b}(2a_{5j}-a_{3j}+a_{5j})=\delta_{5j}$

At nodes of Φ_6: $\dfrac{T_3}{a}(2a_{6j}-a_{5j}-a_{6j})+\dfrac{S_2}{b}(2a_{6j}-a_{4j}+a_{6j})=\delta_{6j}$

that is,

$$
\begin{bmatrix}
2\left(\dfrac{T_1}{a}+\dfrac{S_1}{b}\right) & -\dfrac{T_1}{a} & -\dfrac{S_1}{b} & 0 & 0 & 0 \\
-\dfrac{T_1}{a} & \dfrac{T_1}{a}+\dfrac{2S_2}{b} & 0 & -\dfrac{S_2}{b} & 0 & 0 \\
-\dfrac{S_1}{b} & 0 & 2\left(\dfrac{T_2}{a}+\dfrac{S_1}{b}\right) & -\dfrac{T_2}{a} & -\dfrac{S_1}{b} & 0 \\
0 & -\dfrac{S_2}{b} & -\dfrac{T_2}{a} & \dfrac{T_2}{a}+\dfrac{2S_2}{b} & 0 & -\dfrac{S_2}{b} \\
0 & 0 & -\dfrac{S_1}{b} & 0 & \dfrac{2T_3}{a}+\dfrac{3S_1}{b} & -\dfrac{T_3}{a} \\
0 & 0 & 0 & -\dfrac{S_2}{b} & -\dfrac{T_3}{a} & \dfrac{T_3}{a}+\dfrac{3S_2}{b}
\end{bmatrix}
\begin{bmatrix} a_{1j} \\ a_{2j} \\ a_{3j} \\ a_{4j} \\ a_{5j} \\ a_{6j} \end{bmatrix}
=
\begin{bmatrix} \delta_{1j} \\ \delta_{2j} \\ \delta_{3j} \\ \delta_{4j} \\ \delta_{5j} \\ \delta_{6j} \end{bmatrix}
$$

(9.22)

with δ_{ij} defined as before.

9.5.3.2 Square 16-node configuration

9.5.3.2.1 Subspace S_1

With reference to Figure 9.3(a), application of Φ_j ($j = 1,2,3$) yields the following nodal equations for force equilibrium in the vertical direction:

At nodes of Φ_1: $\dfrac{T_1}{a}(2a_{1j}-0-a_{2j})+\dfrac{T_1}{a}(2a_{1j}-0-a_{2j})=\delta_{1j}$

At nodes of Φ_2: $\dfrac{T_1}{a}(2a_{2j}-a_{1j}-a_{2j})+\dfrac{T_2}{a}(2a_{2j}-0-a_{3j})=\delta_{2j}$

At nodes of Φ_3: $\dfrac{T_2}{a}(2a_{3j} - a_{2j} - a_{3j}) + \dfrac{T_2}{a}(2a_{3j} - a_{2j} - a_{3j}) = \delta_{3j}$

that is,

$$
\begin{bmatrix}
\dfrac{4T_1}{a} & -\dfrac{2T_1}{a} & 0 \\[2ex]
-\dfrac{T_1}{a} & \dfrac{T_1 + 2T_2}{a} & -\dfrac{T_2}{a} \\[2ex]
0 & -\dfrac{2T_2}{a} & \dfrac{2T_2}{a}
\end{bmatrix}
\begin{bmatrix}
a_{1j} \\[1ex]
a_{2j} \\[1ex]
a_{3j}
\end{bmatrix}
=
\begin{bmatrix}
\delta_{1j} \\[1ex]
\delta_{2j} \\[1ex]
\delta_{3j}
\end{bmatrix}
\tag{9.23}
$$

where $\delta_{ij} = 1$ if $i = j$; $\delta_{ij} = 0$ if $i \neq j$.

9.5.3.2.2 Subspace S_2

With reference to Figure 9.3(b), application of Φ_j ($j = 1$) yields the following nodal equation for force equilibrium in the vertical direction:

At nodes of Φ_1: $\dfrac{T_1}{a}(2a_{1j} - 0 + a_{1j}) + \dfrac{T_2}{a}(2a_{1j} - 0 - 0) = \delta_{1j}$

that is,

$$
\left[\dfrac{3T_1 + 2T_2}{a}\right][a_{1j}] = [\delta_{1j}]
\tag{9.24}
$$

with δ_{ij} defined as before (in this case, $\delta_{ij} = \delta_{11} = 1$).

9.5.3.2.3 Subspace S_3

With reference to Figure 9.3(c), application of Φ_j ($j = 1$) yields the following nodal equation for force equilibrium in the vertical direction:

At nodes of Φ_1: $\dfrac{T_1}{a}(2a_{1j} - 0 - a_{1j}) + \dfrac{T_2}{a}(2a_{1j} - 0 - 0) = \delta_{1j}$

that is,

$$
\left[\dfrac{T_1 + 2T_2}{a}\right][a_{1j}] = [\delta_{1j}]
\tag{9.25}
$$

with δ_{ij} defined as before (in this case, $\delta_{ij} = \delta_{11} = 1$).

9.5.3.2.4 Subspace S_4

With reference to Figure 9.3(d), application of Φ_j ($j = 1, 2, 3$) yields the following nodal equations for force equilibrium in the vertical direction:

At nodes of Φ_1: $\dfrac{T_1}{a}(2a_{1j} - 0 - a_{2j}) + \dfrac{T_1}{a}(2a_{1j} - 0 - a_{2j}) = \delta_{1j}$

At nodes of Φ_2: $\dfrac{T_1}{a}(2a_{2j} - a_{1j} + a_{2j}) + \dfrac{T_2}{a}(2a_{2j} - 0 - a_{3j}) = \delta_{2j}$

At nodes of Φ_3: $\dfrac{T_2}{a}(2a_{3j} - a_{2j} + a_{3j}) + \dfrac{T_2}{a}(2a_{3j} - a_{2j} + a_{3j}) = \delta_{3j}$

that is,

$$
\begin{bmatrix}
\dfrac{4T_1}{a} & -\dfrac{2T_1}{a} & 0 \\[2ex]
-\dfrac{T_1}{a} & \dfrac{3T_1 + 2T_2}{a} & -\dfrac{T_2}{a} \\[2ex]
0 & -\dfrac{2T_2}{a} & \dfrac{6T_2}{a}
\end{bmatrix}
\begin{bmatrix}
a_{1j} \\[1ex] a_{2j} \\[1ex] a_{3j}
\end{bmatrix}
=
\begin{bmatrix}
\delta_{1j} \\[1ex] \delta_{2j} \\[1ex] \delta_{3j}
\end{bmatrix}
\tag{9.26}
$$

with δ_{ij} defined as before.

9.5.3.2.5 Subspace S_{51}

With reference to Figure 9.3(e), application of Φ_j' ($j = 1, 2, 3, 4$) yields the following nodal equations for force equilibrium in the vertical direction:

At nodes of Φ_1': $\dfrac{T_1}{a}(2a_{1j} - 0 - a_{4j}) + \dfrac{T_1}{a}(2a_{1j} - 0 - a_{4j}) = \delta_{1j}$

At nodes of Φ_2': $\dfrac{T_2}{a}(2a_{2j} - a_{4j} - 0) + \dfrac{T_2}{a}(2a_{2j} - a_{4j} - 0) = \delta_{2j}$

At nodes of Φ_3': $\dfrac{T_1}{a}(2a_{3j} - a_{4j} - 0) + \dfrac{T_2}{a}(2a_{3j} - 0 - 0) = \delta_{3j}$

At nodes of Φ_4': $\dfrac{T_1}{a}(2a_{4j} - a_{1j} - a_{3j}) + \dfrac{T_2}{a}(2a_{4j} - 0 - a_{2j}) = \delta_{4j}$

that is,

$$
\begin{bmatrix}
\dfrac{4T_1}{a} & 0 & 0 & -\dfrac{2T_1}{a} \\[2ex]
0 & \dfrac{4T_2}{a} & 0 & -\dfrac{2T_2}{a} \\[2ex]
0 & 0 & \dfrac{2(T_1+T_2)}{a} & -\dfrac{T_1}{a} \\[2ex]
-\dfrac{T_1}{a} & -\dfrac{T_2}{a} & -\dfrac{T_1}{a} & \dfrac{2(T_1+T_2)}{a}
\end{bmatrix}
\begin{bmatrix}
a_{1j} \\ a_{2j} \\ a_{3j} \\ a_{4j}
\end{bmatrix}
=
\begin{bmatrix}
\delta_{1j} \\ \delta_{2j} \\ \delta_{3j} \\ \delta_{4j}
\end{bmatrix}
\tag{9.27}
$$

with δ_{ij} defined as before.

9.6 SUBSPACE MASS MATRICES

Let the concentrated masses assumed at the nodes of the two cable nets (such masses may be actual, or lumped parameters of a distributed mass system) be arranged consistently with the general symmetry of the respective cable configurations, as follows:

- For the 24-node C_{2v} configuration, with reference to Figure 9.1(a): m_1 at each position of the nodal set {1, 4, 21, 24}, m_2 at each position of the nodal set {2, 3, 22, 23}, m_3 at each position of the nodal set {5, 8, 17, 20}, m_4 at each position of the nodal set {6, 7, 18, 19}, m_5 at each position of the nodal set {9, 12, 13, 16}, m_6 at each position of the nodal set {10, 11, 14, 15}.
- For the 16-node C_{4v} configuration, with reference to Figure 9.1(b): m_1 at each position of the nodal set {1, 4, 13, 16}, m_2 at each position of the nodal set {2, 3, 5, 8, 9, 12, 14, 15}, m_3 at each position of the nodal set {6, 7, 10, 11}.

As in the examples of previous chapters, the diagonal mass matrix $\mathbf{M}$ for a given subspace consists of nonzero diagonal elements m_{ii} ($i = 1,2,\ldots,r$), which are the values of the mass at each of the nodes of basis vector $\mathbf{\Phi}_i$. Thus, for the 24-node C_{2v} example, $\mathbf{M}$ is the same for all the four subspaces S_1, S_2, S_3 and S_4, and given by

$$
\mathbf{M} =
\begin{bmatrix}
m_1 & 0 & 0 & 0 & 0 & 0 \\
0 & m_2 & 0 & 0 & 0 & 0 \\
0 & 0 & m_3 & 0 & 0 & 0 \\
0 & 0 & 0 & m_4 & 0 & 0 \\
0 & 0 & 0 & 0 & m_5 & 0 \\
0 & 0 & 0 & 0 & 0 & m_6
\end{bmatrix}
\tag{9.28}
$$

For the 16-node C_{4v} example, the mass matrices $\{M_1, M_2, M_3, M_4, M_{51}, M_{52}\}$ for subspaces $\{S_1, S_2, S_3, S_4, S_{51}, S_{52}\}$, respectively, are as follows:

$$M_1 = M_4 = \begin{bmatrix} m_1 & 0 & 0 \\ 0 & m_2 & 0 \\ 0 & 0 & m_3 \end{bmatrix} \tag{9.29}$$

$$M_2 = M_3 = [m_2] \tag{9.30}$$

$$M_{51} = M_{52} = \begin{bmatrix} m_1 & 0 & 0 & 0 \\ 0 & m_3 & 0 & 0 \\ 0 & 0 & m_2 & 0 \\ 0 & 0 & 0 & m_2 \end{bmatrix} \tag{9.31}$$

9.7 EIGENVALUES, EIGENVECTORS AND MODE SHAPES

The system eigenvalues $\lambda (=1/\omega^2$, where ω is a natural circular frequency of the system) are obtained separately for each subspace, from the condition

$$\left| A_l - \lambda M_l^{-1} \right| = 0 \tag{9.32}$$

where A_l, the subspace flexibility matrix, consists of elements a_{ij} ($i = 1, 2, ..., r$; $j = 1, 2, ..., r$) as obtained in Section 9.5, and M_l, the subspace mass matrix, consists of nonzero diagonal elements m_{ii} ($i = 1, 2, ..., r$) as obtained in Section 9.6. Writing out the above determinant in full, we obtain

$$\begin{vmatrix} a_{11} - (\lambda / m_{11}) & a_{12} & \cdot & a_{1r} \\ a_{21} & a_{22} - (\lambda / m_{22}) & \cdot & a_{2r} \\ \cdot & \cdot & \cdot & \cdot \\ a_{r1} & a_{r2} & \cdot & a_{rr} - (\lambda / m_{rr}) \end{vmatrix}_l = 0 \tag{9.33}$$

which may then be expanded into an r th-degree polynomial equation in λ, leading to the r roots for λ that are associated with the subspace in question. The process is repeated for all the subspaces of a problem, in this way generating all the required eigenvalues (hence natural circular frequencies) of the system. As stated in Chapter 6, the eigenvalues yielded from the individual subspaces by the group-theoretic procedure are also the actual eigenvalues of the original problem.

Eigenvectors $\mathbf{U}_l$ are also obtained separately for each subspace, by substituting the now known r eigenvalues of the subspace, one at a time, into the subspace eigenvalue equation

$$\left(\mathbf{A}_l - \lambda \mathbf{M}_l^{-1}\right)\mathbf{U}_l = 0 \qquad (9.34)$$

which, when expanded, becomes

$$
\begin{bmatrix}
a_{11} - (\lambda / m_{11}) & a_{12} & \cdot & a_{1r} \\
a_{21} & a_{22} - (\lambda / m_{22}) & \cdot & a_{2r} \\
\cdot & \cdot & \cdot & \cdot \\
a_{r1} & a_{r2} & \cdot & a_{rr} - (\lambda / m_{rr})
\end{bmatrix}_l
\begin{bmatrix}
u_1 \\
u_2 \\
\cdot \\
u_r
\end{bmatrix}_l
=
\begin{bmatrix}
0 \\
0 \\
\cdot \\
0
\end{bmatrix} \qquad (9.35)
$$

Thus, substituting a particular eigenvalue of the subspace into Equation (9.35), and solving for the components $u_1, u_2, ..., u_r$, we obtain the eigenvector $[u_1\, u_2 \cdot u_r]^{\mathrm{T}}$ corresponding to that eigenvalue, which is an *eigenvector in the r-dimensional subspace*.

Noting that the components $u_1, u_2, ..., u_r$ of such an eigenvector correspond to the basis vectors $\Phi_1, \Phi_2, ..., \Phi_r$, respectively, of the subspace in question, the *eigenvector in the original n-dimensional vector space of the problem* (associated with a particular natural frequency of the system) would then be given by allocating the calculated value of a subspace eigenvector component to *all* the cable nodes associated with the corresponding basis vector, with the signs (positive or negative) of the allocations being in accordance with those of the basis-vector coordinates (as given by Equations (9.3) to (9.6) for the 24-node C_{2v} example, or Equations (9.7) to (9.13) in the case of the 16-node C_{4v} example).

The n components $w_1, w_2, ..., w_n$ of the deflection vector (or mode shape) $\mathbf{W}_i$, corresponding to a system eigenvector $\mathbf{U}_i$, are finally obtained through the relatively simple step

$$\mathbf{W}_i = \mathbf{M}^{-1}\mathbf{U}_i \qquad (9.36)$$

where $\mathbf{M}$ is the conventional diagonal mass matrix of the n degree-of-freedom system. This would then complete the free vibration analysis of the cable nets.

9.8 SUMMARY AND CONCLUDING REMARKS

In this chapter, we have formulated an efficient computational scheme for the linear vibration analysis of high-tension cable nets having orthogonal projections in plan, and assumed to be subject to inertial forces concentrated at the cable intersections. For such a scheme to be applicable, the cable net must possess some symmetry properties, identifiable with one or more symmetry groups. A description of the scheme has been followed by a step-by-step illustration of the computational procedure based on two cable-net configurations.

The technique has consisted in decomposing the n-dimensional vector space of a problem with n degrees of freedom (that is, the n vertical displacement functions describing the motion of n concentrated masses) into a number of independent subspaces each of dimension r_i (where $r_i \ll n$), and then solving for the r_i eigenvalues associated with a given subspace independently of those of the other subspaces. In this way, instead of eventually having to solve an nth-degree polynomial characteristic equation in λ for the n roots of λ (that is, eigenvalues), we need only solve, independently of each other, a series of lower-degree characteristic equations in λ, with considerable savings in computational effort.

Thus, for the example of the 24-node C_{2v} cable-net configuration, such a decomposition results in four independent sets of variables leading to four independent sixth-degree characteristic equations in λ. In the case of the 16-node C_{4v} example, the decomposition leads to the following mutually independent characteristic equations in λ: two third degree, two first degree, and one fourth degree (strictly, two fourth-degree equations result, but these are identical to each other, so that only one of these equations actually needs to be solved to yield four doubly repeating roots). These proportions of vector-space decomposition are exactly representative of all cable nets of rectangular (C_{2v}) and square (C_{4v}) plan projections, if the numbers of x and y cables are such that no nodes (that is, cable intersections) lie in the vertical x and y planes of symmetry.

A feature of the group-theoretic approach is its prediction of the number of normal modes of vibration of a given symmetry type, and of the occurrence of *degenerate normal modes* (that is, normal modes of the same frequency of vibration). The latter are always associated with irreducible representations of dimension greater than 1. The number of linearly independent normal modes of a particular frequency is simply given by the dimension of the corresponding irreducible representation. For instance, with regard to group C_{4v}, the fifth irreducible representation R_5 (of the character table) is two-dimensional; hence, the corresponding subspace S_5 is associated with doubly repeating roots.

For group C_{4v}, a pair of closed-form operators has been given, by means of which the characteristic polynomial associated with the degenerate subspace S_5 is effectively factorized into two identical polynomials, enabling the doubly repeating roots of the subspace to be obtained through the consideration of either of these polynomials.

Once the eigenvalues (which are the same for the system as for the individual subspaces) have been obtained, the associated eigenvectors readily follow: taking one particular eigenvalue (λ) at a time, r eigenvector components $\{u_1, u_2, ..., u_r\}$ are solved for by substituting the eigenvalue into the eigenvalue equation *for the associated r-dimensional subspace*; for this frequency, the system eigenvector can immediately be written down by allocating the calculated u_i ($i = 1, 2, ..., r$) value to *all* the nodes associated with the basis vector Φ_i, in accordance with the signs (positive or negative) of the coordinates of Φ_i.

As a flexibility approach has been adopted, the system eigenvectors finally require to be premultiplied by the inverse of the system diagonal mass matrix, to give the system mode shapes, thereby completing the free vibration analysis of the cable net.

REFERENCES

1. O. Vilnay and P. Rogers. Statical and Dynamical Response of Cable Nets. *International Journal of Solids and Structures*, 1990, 26, 299–312.
2. O. Vilnay. *Cable Nets and Tensegric Shells*. Ellis Horwood, Chichester, 1990.
3. A. Siev. A General Analysis of Prestressed Nets. *Publications of the International Association for Bridge and Structural Engineering*, 1963, 23, 283–292.
4. F. Otto (ed.). *Tensile Structures*. MIT Press, Cambridge, MA, 1966.
5. H.A. Buchholdt, M. Davies and M.J.L. Hussey. The Analysis of Cable Nets. *Journal of the Institute of Mathematics and Its Applications*, 1968, 4, 339–358.
6. H.M. Irvine. *Cable Structures*. MIT Press, Cambridge, MA, 1981.
7. C.R. Calladine. Modal Stiffnesses of a Pretensioned Cable Net. *International Journal of Solids and Structures*, 1982, 18, 829–846.
8. S. Pellegrino and C.R. Calladine. Two-Step Matrix Analysis of Prestressed Cable Nets. In *Proceedings of the Third International Conference on Space Structures*, 1984, 744–749.
9. J. Szabo and L. Kollar. *Structural Design of Cable-Suspended Roofs*. Akademiai Kiado, Budapest, 1984.

Chapter 10

Finite-difference formulations for plates

10.1 GENERAL FINITE-DIFFERENCE FORMULATION FOR PLATE VIBRATION

The vibration of plates represents an important problem in structural mechanics, the solutions being applicable to the design of many structural components such as bridge decks, building floors, platforms carrying machinery, aircraft partitions and vehicle panels. The equation of motion for the undamped free vibration of a rectangular plate may be written as [1, 2]

$$\frac{\partial^4 w}{\partial x^4} + 2\frac{\partial^4 w}{\partial x^2\, \partial y^2} + \frac{\partial^4 w}{\partial y^4} + \frac{\rho}{D}\frac{\partial^2 w}{\partial t^2} = 0 \tag{10.1}$$

where w is the transverse displacement at a point defined by the coordinates $\{x, y\}$ at any given time t, D is the flexural rigidity of the plate and ρ is the mass per unit area of the plate. If the plate has constant thickness h and constant material properties E (Young's modulus of elasticity) and v (Poisson's ratio), then the flexural rigidity is given by

$$D = \frac{Eh^3}{12(1-v^2)} \tag{10.2}$$

Let us assume harmonic vibration. This allows us to express the displacement in the form

$$w(x, y, t) = W(x, y)\, \sin\omega t \tag{10.3}$$

where $W(x, y)$ is a shape function satisfying the boundary conditions and describing the shape of the deflected middle surface of the vibrating plate, and ω is a natural circular frequency of the plate. Substituting for w in Equation (10.1), we obtain

$$\frac{\partial^4 W}{\partial x^4} + 2\frac{\partial^4 W}{\partial x^2\, \partial y^2} + \frac{\partial^4 W}{\partial y^4} - \eta W = 0 \tag{10.4}$$

197

where

$$\eta = \frac{\rho\,\omega^2}{D} \tag{10.5}$$

Various finite-difference schemes for representing differential equations may be seen in standard texts on numerical methods. Adopting the central-difference scheme, and taking equal mesh intervals $d = \Delta x = \Delta y$ in the x and y directions, the finite-difference representation of Equation (10.4) at a pivotal point (m, n) of the mesh is as follows [2]:

$$20W_{m,n} - 8\left(W_{m-1,n} + W_{m+1,n} + W_{m,n-1} + W_{m,n+1}\right)$$

$$+ 2\left(W_{m-1,n-1} + W_{m-1,n+1} + W_{m+1,n-1} + W_{m+1,n+1}\right) \tag{10.6}$$

$$+ W_{m-2,n} + W_{m+2,n} + W_{m,n-2} + W_{m,n+2} - \lambda W_{m,n} = 0$$

where

$$\lambda = \eta\,d^4 \tag{10.7}$$

Figure 10.1 shows the 12 mesh points around a pivotal point denoted by O. Application of Equation (10.6) to this arrangement yields the central-difference equation for point O as

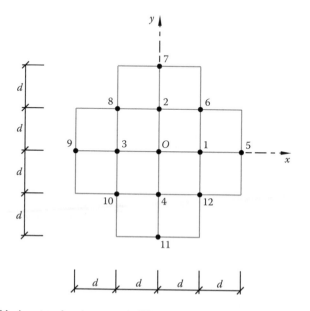

Figure 10.1 Mesh points for the central-difference equation for an interior point O.

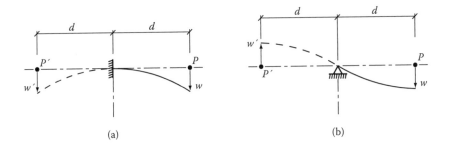

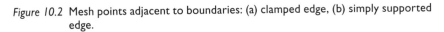

(a)   (b)

Figure 10.2 Mesh points adjacent to boundaries: (a) clamped edge, (b) simply supported
edge.

$$20W_0 - 8(W_1 + W_2 + W_3 + W_4) + 2(W_6 + W_8 + W_{10} + W_{12})$$
$$+ W_5 + W_7 + W_9 + W_{11} - \lambda W_0 = 0$$
(10.8)

For mesh points adjacent to boundaries, it is necessary to introduce what
are called fictitious points. The deflection values to be used for the fictitious
points are obtained by considering the boundary conditions along the edges
of the plate. For instance, the condition of zero slope across a clamped edge
(Figure 10.2(a)) requires the imaginary deflection at the fictitious point to
be the same in magnitude and sign as that at the corresponding real mesh
point, while the condition of zero moment and zero deflection across a simply supported edge (Figure 10.2(b)) requires the imaginary deflection at the
fictitious point to be the same in magnitude but of opposite sign to that at
the corresponding real mesh point.

10.2 GROUP-THEORETIC IMPLEMENTATION

Where a plate has a symmetric configuration, the vibration response will
be expected to exhibit characteristics of the associated symmetry group.
The group-theoretic procedure for the implementation of the finite-difference
method for plate analysis, as recently developed by the author [3], may be
summarized as follows:

1. Identify the symmetry group of the plate configuration. For example,
   rectangular and square plates belong to the symmetry groups C_{2v}
   and C_{4v}, respectively, while regular triangular and hexagonal plates
   belong to the symmetry groups C_{3v} and C_{6v}, respectively.
2. Choose a suitably fine finite-difference mesh whose pattern conforms to the overall symmetry of the plate (e.g. a square grid for a
   square plate, a square or a rectangular grid for a rectangular plate,

a triangular grid for a triangular plate, a triangular or a hexagonal grid for a hexagonal plate). Figure 10.3 illustrates symmetry-conforming finite-difference mesh patterns.

3. Choose the centre of symmetry of the plate as the origin of the coordinate system, plot all symmetry axes of the plate, and label these $x, y, z, 1, 2, 3, a, b, c$ as appropriate. For rectangular and square plates, x and y are the reflection planes of symmetry coinciding with the coordinate axes, while z is the axis of rotational symmetry passing through the origin of the coordinate system and perpendicular to the plate mid-surface. The square plate has the additional reflection planes of symmetry 1 and 2 coinciding with the diagonals of the plate.

4. Number the nodes of the mesh as follows: If a node coincides with the centre of symmetry of the plate, label this as node 1. Choose a node

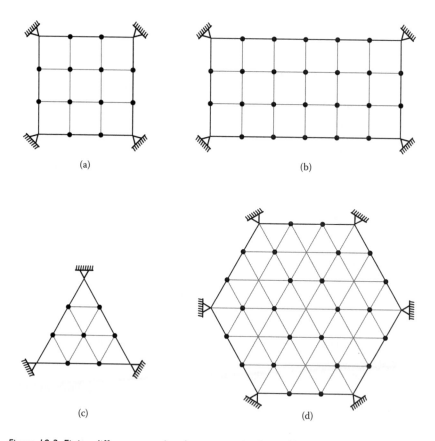

(a)　　　　　　　　　　　　　　　(b)

(c)　　　　　　　　　　　　　　　(d)

Figure 10.3 Finite-difference meshes for symmetric plates: (a) square plate with a square mesh, (b) rectangular plate with a square mesh, (c) triangular plate with a triangular mesh, (d) hexagonal plate with a hexagonal mesh. All plates are simply supported at the corners.

in the positive-positive quadrant of the coordinate system (or on the positive branch of the x or y axis), and nearest to the centre of symmetry, and label this node 2 (this would be node 1 if no node coincides with the centre of symmetry). Then number all nodes belonging to the same permutation set as this node, in the order generated by permuting the node through successive symmetry operations of the group G as given by the character table of the group.

5. Continue the numbering of nodes by picking the next mesh node in the positive-positive quadrant (or on the positive branch of the x or y axis), and number all nodes of the associated permutation set in the same manner as before. Repeat the picking and numbering of nodes (generally moving outward from the centre of symmetry) until all nodes of the mesh are numbered from 1 up to n.

6. For each subspace of the problem, apply the associated idempotent to each nodal function $\phi_1, \phi_2, ..., \phi_n$, and select the independent basis vectors for the subspace.

7. Write down the conventional finite-difference equations *for only the nodes corresponding to the first components of the basis vectors of each subspace* (these are usually the first nodes of the nodal sets of the mesh). The first components of the basis vectors of the various subspaces are usually the same, allowing one set of conventional finite-difference equations to serve for all the subspaces of the problem.

8. For each subspace of the problem, use the associated basis vectors to transform the conventional finite-difference equations (for the nodes corresponding to the first components of the basis vectors of the subspace) into symmetry-adapted equations for the subspace.

9. Use the vanishing condition for the determinant of the subspace eigenvalue matrices to solve for the eigenvalues of the various subspaces. These are also the eigenvalues of the conventional finite-difference system. Thus, all natural frequencies of the system are obtained by simply solving for the eigenvalues of the individual subspaces.

10.3 APPLICATION TO RECTANGULAR AND SQUARE PLATES

Figure 10.4(a) shows a rectangular plate simply supported on all four edges, with a regular grid of mesh lines in the x and y directions giving a total of 24 mesh points on the plate. Figure 10.4(b) shows a square plate clamped on all four edges, with a regular grid of mesh lines in the x and y directions giving a total of 25 mesh points on the plate. In both cases, the origin of the x, y coordinate system is located at the centre of symmetry of the plate. In the case of the square plate, the additional diagonal planes of symmetry are labelled 1–1 and 2–2. These two examples are taken from reference [3].

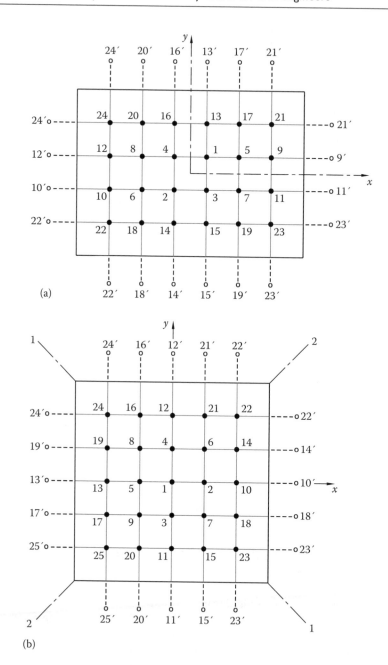

Figure 10.4 Illustrative examples: (a) rectangular plate simply supported along all four edges and having a square mesh with 24 mesh points, (b) square plate clamped along all four edges and having a square mesh with 25 mesh points.

Clearly, the mesh pattern for the rectangular plate conforms to symmetry group C_{2v}, while the mesh pattern for the square plate conforms to symmetry group C_{4v}. Mesh points on the plates have been numbered in accordance with the procedure outlined in the previous section. Relevant fictitious nodes are shown outside the boundary of the plate, as mirror images of the adjacent real nodes on the plate.

10.3.1 Basis vectors for the rectangular plate

In accordance with step 6 of the previous section, we apply the relevant idempotent of the symmetry group C_{2v} to each of the 24 nodal functions $\phi_1, \phi_2, ..., \phi_{24}$ associated with the mesh points of the rectangular plate. The idempotents $P^{(i)}$ for subspaces $S^{(i)}$ ($i = 1, ..., 4$) were given in Chapter 6 (here, we choose to show the i as a superscript in brackets). This results in 24 linear combinations of the nodal functions for each subspace, not all of which are independent. We then select a set of linearly independent combinations of functions as the basis vectors for the subspace in question. It is found that all four subspaces are six-dimensional (i.e. have six basis vectors each). The basis vectors for the four subspaces are listed below. *In this chapter and the next, we will adopt a rather more elaborate notation for basis vectors, where the parent subspace is denoted by a numerical superscript in brackets, and the basis-vector number by a numerical subscript.*

Subspace $S^{(1)}$

$$\Phi_1^{(1)} = \phi_1 + \phi_2 + \phi_3 + \phi_4 \tag{10.9a}$$

$$\Phi_2^{(1)} = \phi_5 + \phi_6 + \phi_7 + \phi_8 \tag{10.9b}$$

$$\Phi_3^{(1)} = \phi_9 + \phi_{10} + \phi_{11} + \phi_{12} \tag{10.9c}$$

$$\Phi_4^{(1)} = \phi_{13} + \phi_{14} + \phi_{15} + \phi_{16} \tag{10.9d}$$

$$\Phi_5^{(1)} = \phi_{17} + \phi_{18} + \phi_{19} + \phi_{20} \tag{10.9e}$$

$$\Phi_6^{(1)} = \phi_{21} + \phi_{22} + \phi_{23} + \phi_{24} \tag{10.9f}$$

Subspace $S^{(2)}$

$$\Phi_1^{(2)} = \phi_1 + \phi_2 - \phi_3 - \phi_4 \tag{10.10a}$$

$$\Phi_2^{(2)} = \phi_5 + \phi_6 - \phi_7 - \phi_8 \tag{10.10b}$$

$$\Phi_3^{(2)} = \phi_9 + \phi_{10} - \phi_{11} - \phi_{12} \tag{10.10c}$$

$$\Phi_4^{(2)} = \phi_{13} + \phi_{14} - \phi_{15} - \phi_{16} \tag{10.10d}$$

$$\Phi_5^{(2)} = \phi_{17} + \phi_{18} - \phi_{19} - \phi_{20} \tag{10.10e}$$

$$\Phi_6^{(2)} = \phi_{21} + \phi_{22} - \phi_{23} - \phi_{24} \tag{10.10f}$$

Subspace $S^{(3)}$

$$\Phi_1^{(3)} = \phi_1 - \phi_2 + \phi_3 - \phi_4 \tag{10.11a}$$

$$\Phi_2^{(3)} = \phi_5 - \phi_6 + \phi_7 - \phi_8 \tag{10.11b}$$

$$\Phi_3^{(3)} = \phi_9 - \phi_{10} + \phi_{11} - \phi_{12} \tag{10.11c}$$

$$\Phi_4^{(3)} = \phi_{13} - \phi_{14} + \phi_{15} - \phi_{16} \tag{10.11d}$$

$$\Phi_5^{(3)} = \phi_{17} - \phi_{18} + \phi_{19} - \phi_{20} \tag{10.11e}$$

$$\Phi_6^{(3)} = \phi_{21} - \phi_{22} + \phi_{23} - \phi_{24} \tag{10.11f}$$

Subspace $S^{(4)}$

$$\Phi_1^{(4)} = \phi_1 - \phi_2 - \phi_3 + \phi_4 \tag{10.12a}$$

$$\Phi_2^{(4)} = \phi_5 - \phi_6 - \phi_7 + \phi_8 \tag{10.12b}$$

$$\Phi_3^{(4)} = \phi_9 - \phi_{10} - \phi_{11} + \phi_{12} \tag{10.12c}$$

$$\Phi_4^{(4)} = \phi_{13} - \phi_{14} - \phi_{15} + \phi_{16} \tag{10.12d}$$

$$\Phi_5^{(4)} = \phi_{17} - \phi_{18} - \phi_{19} + \phi_{20} \tag{10.12e}$$

$$\Phi_6^{(4)} = \phi_{21} - \phi_{22} - \phi_{23} + \phi_{24} \tag{10.12f}$$

We may visualize the symmetry associated with the above subspaces by reference to Figure 10.5, where the sixth basis vector of each subspace has been plotted. Only one basis vector need be plotted for each subspace, since basis vectors of the same subspace all have the same symmetry.

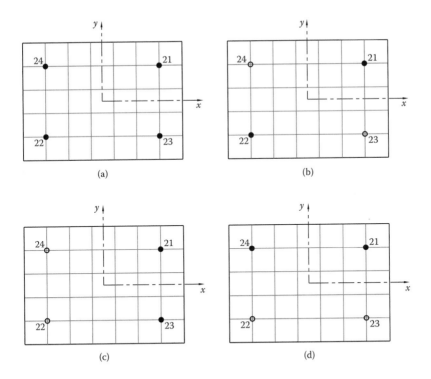

Figure 10.5 Symmetries of the subspaces of the rectangular plate illustrated by the sixth basis vector of each subspace: (a) $\Phi_6^{(1)}$ of subspace $S^{(1)}$, (b) $\Phi_6^{(2)}$ of subspace $S^{(2)}$, (c) $\Phi_6^{(3)}$ of subspace $S^{(3)}$, (d) $\Phi_6^{(4)}$ of subspace $S^{(4)}$. Black dots denote positive coordinates of the basis vectors, while rings denote negative coordinates.

10.3.2 Basis vectors for the square plate

We implement step 6 of the previous section by applying the idempotents $P^{(1)}$, $P^{(2)}$, $P^{(3)}$, $P^{(4)}$, $P^{(5,1)}$ and $P^{(5,2)}$ of the symmetry group C_{4v} (corresponding to subspaces $S^{(1)}$, $S^{(2)}$, $S^{(3)}$, $S^{(4)}$, $S^{(5,1)}$ and $S^{(5,2)}$, respectively) to each of the 25 nodal functions $\phi_1, \phi_2, \ldots, \phi_{25}$ associated with the mesh points of the square plate. The idempotents for symmetry group C_{4v} were given earlier as Equations (6.4) and the associated special operators of Equations (9.14). For each subspace, this results in 25 linear combinations of the ϕ functions, not all of which are independent. Selecting a set of linearly independent combinations of ϕ functions as the basis vectors for the subspace, we obtain the following results:

Subspace $S^{(1)}$

$$\Phi_1^{(1)} = \phi_1 \tag{10.13a}$$

$$\Phi_2^{(1)} = \phi_2 + \phi_3 + \phi_4 + \phi_5 \tag{10.13b}$$

$$\Phi_3^{(1)} = \phi_6 + \phi_7 + \phi_8 + \phi_9 \tag{10.13c}$$

$$\Phi_4^{(1)} = \phi_{10} + \phi_{11} + \phi_{12} + \phi_{13} \tag{10.13d}$$

$$\Phi_5^{(1)} = \phi_{14} + \phi_{15} + \phi_{16} + \phi_{17} + \phi_{18} + \phi_{19} + \phi_{20} + \phi_{21} \tag{10.13e}$$

$$\Phi_6^{(1)} = \phi_{22} + \phi_{23} + \phi_{24} + \phi_{25} \tag{10.13f}$$

Subspace $S^{(2)}$

$$\Phi_1^{(2)} = \phi_{14} + \phi_{15} + \phi_{16} + \phi_{17} - \phi_{18} - \phi_{19} - \phi_{20} - \phi_{21} \tag{10.14}$$

Subspace $S^{(3)}$

$$\Phi_1^{(3)} = \phi_2 - \phi_3 - \phi_4 + \phi_5 \tag{10.15a}$$

$$\Phi_2^{(3)} = \phi_{10} - \phi_{11} - \phi_{12} + \phi_{13} \tag{10.15b}$$

$$\Phi_3^{(3)} = \phi_{14} - \phi_{15} - \phi_{16} + \phi_{17} + \phi_{18} + \phi_{19} - \phi_{20} - \phi_{21} \tag{10.15c}$$

Subspace $S^{(4)}$

$$\Phi_1^{(4)} = \phi_6 - \phi_7 - \phi_8 + \phi_9 \tag{10.16a}$$

$$\Phi_2^{(4)} = \phi_{14} - \phi_{15} - \phi_{16} + \phi_{17} - \phi_{18} - \phi_{19} + \phi_{20} + \phi_{21} \tag{10.16b}$$

$$\Phi_3^{(4)} = \phi_{22} - \phi_{23} - \phi_{24} + \phi_{25} \tag{10.16c}$$

Subspace $S^{(5,1)}$

$$\Phi_1^{(5,1)} = \phi_2 + \phi_3 - \phi_4 - \phi_5 \tag{10.17a}$$

$$\Phi_2^{(5,1)} = \phi_7 - \phi_8 \tag{10.17b}$$

$$\Phi_3^{(5,1)} = \phi_{10} + \phi_{11} - \phi_{12} - \phi_{13} \tag{10.17c}$$

$$\Phi_4^{(5,1)} = \phi_{14} - \phi_{17} + \phi_{20} - \phi_{21} \tag{10.17d}$$

$$\Phi_5^{(5,1)} = \phi_{15} - \phi_{16} + \phi_{18} - \phi_{19} \tag{10.17e}$$

$$\Phi_6^{(5,1)} = \phi_{23} - \phi_{24} \tag{10.17f}$$

Subspace $S^{(5,2)}$

$$\Phi_1^{(5,2)} = \phi_2 - \phi_3 + \phi_4 - \phi_5 \tag{10.18a}$$

$$\Phi_2^{(5,2)} = \phi_6 - \phi_9 \tag{10.18b}$$

$$\Phi_3^{(5,2)} = \phi_{10} - \phi_{11} + \phi_{12} - \phi_{13} \tag{10.18c}$$

$$\Phi_4^{(5,2)} = \phi_{14} - \phi_{17} - \phi_{20} + \phi_{21} \tag{10.18d}$$

$$\Phi_5^{(5,2)} = \phi_{15} - \phi_{16} - \phi_{18} + \phi_{19} \tag{10.18e}$$

$$\Phi_6^{(5,2)} = \phi_{22} - \phi_{25} \tag{10.18f}$$

The symmetries associated with the first four subspaces ($S^{(1)}$ to $S^{(4)}$) are depicted in Figure 10.6, where only one representative basis vector of each subspace has been selected for plotting. For subspaces $S^{(5,1)}$ and $S^{(5,2)}$, we have plotted all the basis vectors of each subspace (Figure 10.7), not only to show the symmetries associated with these subspaces, but more importantly, to illustrate the fact that subspaces $S^{(5,1)}$ and $S^{(5,2)}$ always yield sets of basis vectors that are identical except for orientation.

The plots for subspace $S^{(5,1)}$ are symmetrical about the diagonal axis 1-1 and antisymmetrical about the diagonal 2-2 axis, while those for subspace $S^{(5,2)}$ are symmetrical about the axis 2-2 and antisymmetrical about the 1-1 axis. As pointed out in Chapter 9, orientation of the basis vectors does not affect the physical properties of the system (frequencies and mode shapes). Subspaces $S^{(5,1)}$ and $S^{(5,2)}$ yield identical sets of solutions for the eigenvalues, so it will be sufficient to consider only one of these subspaces. The matching solutions of subspaces $S^{(5,1)}$ and $S^{(5,2)}$ correspond, of course, to doubly repeating roots in the full space of the conventional problem.

10.3.3 Nodal sets

Let us define a nodal set of a finite-difference mesh as the set of all nodes that are permuted by the symmetry elements of the group G of the system. For instance, considering the rectangular plate (Figure 10.4(a)), if we apply the symmetry elements $\{e, C_2, \sigma_x, \sigma_y\}$ of the group C_{2v} to nodal position 1

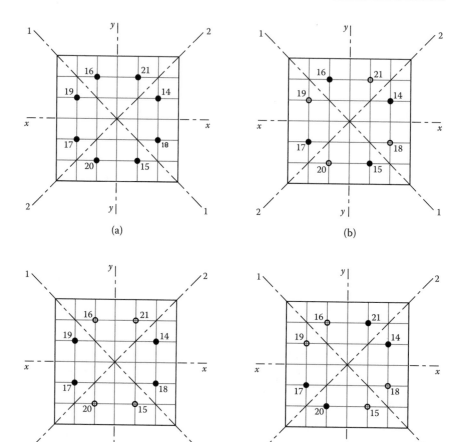

Figure 10.6 Symmetries of the first four subspaces of the square plate illustrated by one basis vector of each subspace: (a) $\Phi_5^{(1)}$ of subspace $S^{(1)}$, (b) $\Phi_1^{(2)}$ of subspace $S^{(2)}$, (c) $\Phi_3^{(3)}$ of subspace $S^{(3)}$, (d) $\Phi_2^{(4)}$ of subspace $S^{(4)}$. Black dots denote positive coordinates of the basis vectors, while rings denote negative coordinates.

of the mesh, we generate the nodal positions $\{1,2,3,4\}$, which therefore constitute a nodal set of the system. Nodal sets of the finite-difference mesh of the square plate (Figure 10.4(b)) are the sets of all positions that are permuted by the symmetry elements $\{e, C_4, C_4^{-1}, C_2, \sigma_x, \sigma_y, \sigma_1, \sigma_2\}$ of the group C_{4v}. Nodal sets for the two plates are therefore:

Rectangular plate: $\{1, 2, 3, 4\}$

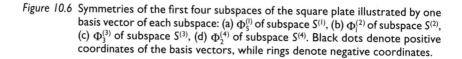

$\{5, 6, 7, 8\}$

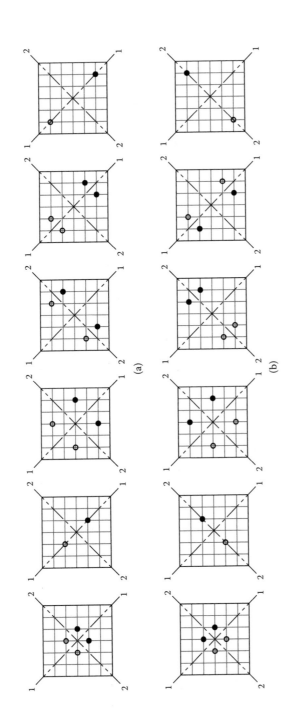

(a)

(b)

Figure 10.7 Basis vectors of subspaces $S^{(5,1)}$ and $S^{(5,2)}$ of the square plate, and their symmetries: (a) subspace $S^{(5,1)}$: $\left\{ \Phi_1^{(5,1)}, \Phi_2^{(5,1)}, \Phi_3^{(5,1)}, \Phi_4^{(5,1)}, \Phi_5^{(5,1)}, \Phi_6^{(5,1)} \right\}$, (b) subspace $S^{(5,2)}$: $\left\{ \Phi_1^{(5,2)}, \Phi_2^{(5,2)}, \Phi_3^{(5,2)}, \Phi_4^{(5,2)}, \Phi_5^{(5,2)}, \Phi_6^{(5,2)} \right\}$.

$$\{9, 10, 11, 12\}$$

$$\{13, 14, 15, 16\}$$

$$\{17, 18, 19, 20\}$$

$$\{21, 22, 23, 24\}$$

Square plate: $\{1\}$

$$\{2, 3, 4, 5\}$$

$$\{6, 7, 8, 9\}$$

$$\{10, 11, 12, 13\}$$

$$\{14, 15, 16, 17, 18, 19, 20, 21\}$$

$$\{22, 23, 24, 25\}$$

The consecutive numbering within the nodal sets is a direct consequence of the rules for node numbering that we adopted in steps 4 and 5 of Section 10.2.

10.4 FINITE-DIFFERENCE EQUATIONS FOR GENERATOR NODES OF THE BASIS VECTORS

10.4.1 Generator nodes

In the present approach, it is only necessary to write down the finite-difference equations for the nodes corresponding to the first components of the basis vectors of each subspace, and operate on this reduced set of equations in order to generate all the required symmetry-adapted finite-difference equations for the various subspaces of the problem. We will refer to these special nodes as generator nodes of the basis vectors.

Considering the rectangular plate (C_{2v} symmetry), all subspaces $S^{(i)}$ ($i = 1$, 2, 3, 4) each have six basis vectors $\left\{ \Phi_1^{(i)}, \Phi_2^{(i)}, \Phi_3^{(i)}, \Phi_4^{(i)}, \Phi_5^{(i)}, \Phi_6^{(i)} \right\}$ whose first components are $\left\{ \phi_1, \phi_5, \phi_9, \phi_{13}, \phi_{17}, \phi_{21} \right\}$, respectively, irrespective of the subspace (see Equations (10.9) to (10.12)). The nodes corresponding to the first components of the basis vectors (that is, the generator nodes) are therefore $\{1, 5, 9, 13, 17, 21\}$ for all four subspaces. We note that these are also the first nodes of the nodal sets given in Section 10.3.3. We will therefore need to write down finite-difference relationships for these 6 nodes only (not for the entire 24 nodes of the problem), and be able to derive reduced sets of equations for *all* subspaces on the basis of these.

In the case of the square plate (C_{4v} symmetry), the number of basis vectors spanning the subspace (that is, the dimension of the subspace) is no longer the same for all the subspaces. The first components of the basis vectors $\Phi_1^{(i)}, \Phi_2^{(i)}, ..., \Phi_r^{(i)}$, for each subspace $S^{(i)}$ of dimension r, are as follows (refer to Equations (10.13) to (10.18)):

Subspace $S^{(1)}$: $\{\phi_1, \phi_2, \phi_6, \phi_{10}, \phi_{14}, \phi_{22}\}$

Subspace $S^{(2)}$: $\{\phi_{14}\}$

Subspace $S^{(3)}$: $\{\phi_2, \phi_{10}, \phi_{14}\}$

Subspace $S^{(4)}$: $\{\phi_6, \phi_{14}, \phi_{22}\}$

Subspace $S^{(5,1)}$: $\{\phi_2, \phi_7, \phi_{10}, \phi_{14}, \phi_{15}, \phi_{23}\}$

Subspace $S^{(5,2)}$: $\{\phi_2, \phi_6, \phi_{10}, \phi_{14}, \phi_{15}, \phi_{22}\}$

For subspace $S^{(1)}$, the generator nodes of the basis vectors are therefore {1, 2, 6, 10, 14, 22}, which are also the first nodes of the nodal sets given in Section 10.3.3. The sets of generator nodes for subspaces $S^{(2)}$, $S^{(3)}$ and $S^{(4)}$ are all subsets of the set of generator nodes for subspace $S^{(1)}$, so the set of finite-difference equations written down for subspace $S^{(1)}$ will also cover subspace $S^{(2)}$ (where we will need the equation for node 14 only), subspace $S^{(3)}$ (where we will need the equations for nodes 2, 10 and 14 only) and subspace $S^{(4)}$ (where we will need the equations for nodes 6, 14 and 22 only).

Subspaces $S^{(5,1)}$ and $S^{(5,2)}$ need to be treated a little differently. Subspace $S^{(5,1)}$ will require the finite-difference equations for nodes 2, 10 and 14, which are already included in the set of equations for subspace $S^{(1)}$, and the additional finite-difference equations for nodes 7, 15 and 23. On the other hand, subspace $S^{(5,2)}$ will require the finite-difference equations for nodes 2, 6, 10, 14 and 22 (all of which are already included in the set of equations for subspace $S^{(1)}$), and only one additional finite-difference equation for node 15. Since subspaces $S^{(5,1)}$ and $S^{(5,2)}$ yield identical sets of eigenvalues for reasons already explained, we need consider only one of them. We choose subspace $S^{(5,2)}$ since this requires only one additional finite-difference equation (for node 15).

Summarizing, the nodes for which finite-difference equations need to be written down are {1, 5, 9, 13, 17, 21} for the rectangular plate and {1, 2, 6, 10, 14, 15, 22} for the square plate.

10.4.2 Basis finite-difference equations for the rectangular plate

Since all edges are simply supported, $W = 0$ for all nodes lying on the edges of the plate. For the fictitious nodes of the finite-difference mesh (refer to Figure 10.4(a)), the deflections W are equal in magnitude but of opposite sign to those of the corresponding real nodes. The finite-difference equations for the relevant nodes of the problem (i.e. nodes {1, 5, 9, 13, 17, 21}) are as follows:

Node 1

$$(20 - \lambda)W_1 + 2W_2 - 8W_3 - 8W_4 - 8W_5 + 2W_7 + W_8 + W_9$$
$$- 8W_{13} + W_{15} + 2W_{16} + 2W_{17} = 0 \tag{10.19a}$$

Node 5

$$-8W_1 + 2W_3 + W_4 + (20 - \lambda)W_5 - 8W_7 - 8W_9 + 2W_{11} + 2W_{13}$$
$$- 8W_{17} + W_{19} + 2W_{21} = 0 \tag{10.19b}$$

Node 9

$$W_1 - 8W_5 + 2W_7 + (19 - \lambda)W_9 - 8W_{11} + 2W_{17} - 8W_{21} + W_{23} = 0 \tag{10.19c}$$

Node 13

$$-8W_1 + W_3 + 2W_4 + 2W_5 + (19 - \lambda)W_{13} - 8W_{16} - 8W_{17} + W_{20} + W_{21} = 0 \tag{10.19d}$$

Node 17

$$2W_1 - 8W_5 + W_7 + 2W_9 - 8W_{13} + W_{16} + (19 - \lambda)W_{17} - 8W_{21} = 0 \tag{10.19e}$$

Node 21

$$2W_5 - 8W_9 + W_{11} + W_{13} - 8W_{17} + (18 - \lambda)W_{21} = 0 \qquad (10.19\text{f})$$

10.4.3 Basis finite-difference equations for the square plate

Since all edges are fixed, $W = 0$ for all nodes lying on the edges of the plate. For the fictitious nodes of the finite-difference mesh (refer to Figure 10.4(b)), the deflections W are equal in magnitude and of the same sign as those of the corresponding real nodes. The finite-difference equations for the relevant nodes of the problem (i.e. nodes $\{1, 2, 6, 10, 14, 15, 22\}$) are

Node 1

$$(20 - \lambda)W_1 - 8W_2 - 8W_3 - 8W_4 - 8W_5 + 2W_6 + 2W_7 + 2W_8$$
$$+ 2W_9 + W_{10} + W_{11} + W_{12} + W_{13} = 0 \qquad (10.20\text{a})$$

Node 2

$$-8W_1 + (20 - \lambda)W_2 + 2W_3 + 2W_4 + W_5 - 8W_6 - 8W_7 - 8W_{10}$$
$$+ 2W_{14} + W_{15} + 2W_{18} + W_{21} = 0 \qquad (10.20\text{b})$$

Node 6

$$2W_1 - 8W_2 - 8W_4 + (20 - \lambda)W_6 + W_7 + W_8 + 2W_{10} + 2W_{12}$$
$$- 8W_{14} - 8W_{21} + 2W_{22} = 0 \qquad (10.20\text{c})$$

Node 10

$$W_1 - 8W_2 + 2W_6 + 2W_7 + (21 - \lambda)W_{10} - 8W_{14} - 8W_{18} + W_{22} + W_{23} = 0$$
$$(10.20\text{d})$$

Node 14

$$2W_2 + W_4 - 8W_6 - 8W_{10} + (21 - \lambda)W_{14} + W_{18} + 2W_{21} - 8W_{22} = 0$$
$$(10.20\text{e})$$

Node 15

$$W_2 + 2W_3 - 8W_7 - 8W_{11} + (21 - \lambda)W_{15} + 2W_{18} + W_{20} - 8W_{23} = 0$$

(10.20f)

Node 22

$$2W_6 + W_{10} + W_{12} - 8W_{14} - 8W_{21} + (22 - \lambda)W_{22} = 0 \qquad (10.20g)$$

10.5 SYMMETRY-ADAPTED FINITE-DIFFERENCE EQUATIONS AND SYSTEM EIGENVALUES

For any given basis vector $\Phi_j^{(1)}$ of subspace $S^{(i)}$, the coefficients of the components ϕ are either +1 or −1 for all subspaces of the C_{2v} problem (rectangular plate) and the C_{4v} problem (square plate). These coefficients give the relative values of the transverse displacements associated with the nodes of Φ_j.

For a given r-dimensional subspace, we will denote the amplitude of the displacements associated with the nodes of Φ_j ($j = 1,2,...,r$) by the parameter f_j. This amplitude will be the same for all nodes of Φ_j. This formulation results in an r-dimensional eigenvalue problem within the subspace $S^{(i)}$, which upon solving yields the r eigenvalues (or natural circular frequencies of vibration of the plate) for that subspace.

Very significantly, these subspace eigenvalues are also eigenvalues of the full vector space of the vibration problem (no further computations are required), and by collecting together all the sets of eigenvalues yielded by the various subspaces of the problem, we obtain the full set of natural frequencies for the plate. The detailed computations for the two examples follow.

10.5.1 Rectangular plate

10.5.1.1 Subspace $S^{(1)}$

From the coefficients of the $\Phi_j^{(1)}$ ($j = 1,2,...,6$), let

$$W_1 = W_2 = W_3 = W_4 = f_1$$

$$W_5 = W_6 = W_7 = W_8 = f_2$$

$$W_9 = W_{10} = W_{11} = W_{12} = f_3$$

$$W_{13} = W_{14} = W_{15} = W_{16} = f_4$$

$$W_{17} = W_{18} = W_{19} = W_{20} = f_5$$

$$W_{21} = W_{22} = W_{23} = W_{24} = f_6$$

Making the above substitutions into each of Equations (10.19a) to (10.19f), we obtain six equations in $\{f_1, f_2, f_3, f_4, f_5, f_6\}$:

$$\begin{bmatrix} (6-\lambda) & -5 & 1 & -5 & 2 & 0 \\ -5 & (12-\lambda) & -6 & 2 & -7 & 2 \\ 1 & -6 & (11-\lambda) & 0 & 2 & -7 \\ -5 & 2 & 0 & (11-\lambda) & -7 & 1 \\ 2 & -7 & 2 & -7 & (19-\lambda) & -8 \\ 0 & 2 & -7 & 1 & -8 & (18-\lambda) \end{bmatrix} \begin{bmatrix} f_1 \\ f_2 \\ f_3 \\ f_4 \\ f_5 \\ f_6 \end{bmatrix} = \begin{bmatrix} 0 \\ 0 \\ 0 \\ 0 \\ 0 \\ 0 \end{bmatrix}$$

$$(10.21)$$

The vanishing condition for the determinant of the above 6×6 matrix yields a sixth-degree polynomial equation in λ, whose roots (the required eigenvalues) are

$$\lambda_1 = 0.336; \lambda_2 = 3.752; \lambda_3 = 7.930; \lambda_4 = 13.169; \lambda_5 = 17.414; \lambda_6 = 34.398$$

10.5.1.2 Subspace $S^{(2)}$

From the coefficients of the $\Phi_j^{(2)}$ $(j = 1, 2, ..., 6)$, let

$$W_1 = W_2 = -W_3 = -W_4 = f_1$$

$$W_5 = W_6 = -W_7 = -W_8 = f_2$$

$$W_9 = W_{10} = -W_{11} = -W_{12} = f_3$$

$$W_{13} = W_{14} = -W_{15} = -W_{16} = f_4$$

$$W_{17} = W_{18} = -W_{19} = -W_{20} = f_5$$

$$W_{21} = W_{22} = -W_{23} = -W_{24} = f_6$$

Making the above substitutions into each of Equations (10.19a) to (10.19f), we obtain

$$
\begin{bmatrix}
(38-\lambda) & -11 & 1 & -11 & 2 & 0 \\
-11 & (28-\lambda) & -10 & 2 & -9 & 2 \\
1 & -10 & (27-\lambda) & 0 & 2 & -9 \\
-11 & 2 & 0 & (27-\lambda) & -9 & 1 \\
2 & -9 & 2 & -9 & (19-\lambda) & -8 \\
0 & 2 & -9 & 1 & -8 & (18-\lambda)
\end{bmatrix}
\begin{bmatrix} f_1 \\ f_2 \\ f_3 \\ f_4 \\ f_5 \\ f_6 \end{bmatrix}
=
\begin{bmatrix} 0 \\ 0 \\ 0 \\ 0 \\ 0 \\ 0 \end{bmatrix}
$$

$$(10.22)$$

The vanishing condition for the determinant of the above 6 × 6 matrix yields a sixth-degree polynomial equation in λ, whose roots (the required eigenvalues) are obtained as

$\lambda_1 = 4.558;\ \lambda_2 = 14.646;\ \lambda_3 = 19.106;\ \lambda_4 = 26.873;$

$\lambda_5 = 36.761;\ \lambda_6 = 55.056$

10.5.1.3 Subspace $S^{(3)}$

From the coefficients of the $\Phi_j^{(3)}$ ($j = 1, 2, ..., 6$), let

$W_1 = -W_2 = W_3 = -W_4 = f_1$

$W_5 = -W_6 = W_7 = -W_8 = f_2$

$W_9 = -W_{10} = W_{11} = -W_{12} = f_3$

$W_{13} = -W_{14} = W_{15} = -W_{16} = f_4$

$W_{17} = -W_{18} = W_{19} = -W_{20} = f_5$

$W_{21} = -W_{22} = W_{23} = -W_{24} = f_6$

Making the above substitutions into each of Equations (10.19a) to (10.19f), we obtain the following six equations in $\{f_1, f_2, f_3, f_4, f_5, f_6\}$:

$$
\begin{bmatrix}
(18-\lambda) & -7 & 1 & -9 & 2 & 0 \\
-7 & (12-\lambda) & -6 & 2 & -7 & 2 \\
1 & -6 & (11-\lambda) & 0 & 2 & -7 \\
-9 & 2 & 0 & (27-\lambda) & -9 & 1 \\
2 & -7 & 2 & -9 & (19-\lambda) & -8 \\
0 & 2 & -7 & 1 & -8 & (18-\lambda)
\end{bmatrix}
\begin{bmatrix}
f_1 \\ f_2 \\ f_3 \\ f_4 \\ f_5 \\ f_6
\end{bmatrix}
=
\begin{bmatrix}
0 \\ 0 \\ 0 \\ 0 \\ 0 \\ 0
\end{bmatrix}
$$

$$(10.23)$$

The vanishing condition for the determinant of the above 6 × 6 matrix yields a sixth-degree polynomial equation in λ, whose roots (the required eigenvalues) are as follows:

$$\lambda_1 = 1.288;\ \lambda_2 = 7.992;\ \lambda_3 = 11.364;\ \lambda_4 = 17.505;\ \lambda_5 = 25.635;\ \lambda_6 = 41.216$$

10.5.1.4 Subspace $S^{(4)}$

From the coefficients of the $\Phi_j^{(4)}$ (j = 1, 2,..., 6), let

$$W_1 = -W_2 = -W_3 = W_4 = f_1$$

$$W_5 = -W_6 = -W_7 = W_8 = f_2$$

$$W_9 = -W_{10} = -W_{11} = W_{12} = f_3$$

$$W_{13} = -W_{14} = -W_{15} = W_{16} = f_4$$

$$W_{17} = -W_{18} = -W_{19} = W_{20} = f_5$$

$$W_{21} = -W_{22} = -W_{23} = W_{24} = f_6$$

Making the above substitutions into each of Equations (10.19a) to (10.19f), we obtain

$$
\begin{bmatrix}
(18-\lambda) & -9 & 1 & -7 & 2 & 0 \\
-9 & (28-\lambda) & -10 & 2 & -9 & 2 \\
1 & -10 & (27-\lambda) & 0 & 2 & -9 \\
-7 & 2 & 0 & (11-\lambda) & -7 & 1 \\
2 & -9 & 2 & -7 & (19-\lambda) & -8 \\
0 & 2 & -9 & 1 & -8 & (18-\lambda)
\end{bmatrix}
\begin{bmatrix}
f_1 \\ f_2 \\ f_3 \\ f_4 \\ f_5 \\ f_6
\end{bmatrix}
=
\begin{bmatrix}
0 \\ 0 \\ 0 \\ 0 \\ 0 \\ 0
\end{bmatrix}
$$

$$(10.24)$$

The vanishing condition for the determinant of the above 6×6 matrix yields a sixth-degree polynomial equation in λ, whose roots (the required eigenvalues) are

$$\lambda_1 = 2.496; \ \lambda_2 = 8.626; \ \lambda_3 = 14.563; \ \lambda_4 = 21.427;$$

$$\lambda_5 = 26.760; \ \lambda_6 = 47.128$$

10.5.2 Square plate

10.5.2.1 Subspace $S^{(1)}$

From the coefficients of the $\Phi_j^{(1)}$ $(j = 1, 2, ..., 6)$, let

$$W_1 = f_1$$

$$W_2 = W_3 = W_4 = W_5 = f_2$$

$$W_6 = W_7 = W_8 = W_9 = f_3$$

$$W_{10} = W_{11} = W_{12} = W_{13} = f_4$$

$$W_{14} = W_{15} = W_{16} = W_{17} = W_{18} = W_{19} = W_{20} = W_{21} = f_5$$

$$W_{22} = W_{23} = W_{24} = W_{25} = f_6$$

Making the above substitutions into each of Equations (10.20a) to (10.20e) and (10.20g), we obtain six equations:

$$\begin{bmatrix} (20-\lambda) & -32 & 8 & 4 & 0 & 0 \\ -8 & (25-\lambda) & -16 & -8 & 6 & 0 \\ 2 & -16 & (22-\lambda) & 4 & -16 & 2 \\ 1 & -8 & 4 & (21-\lambda) & -16 & 2 \\ 0 & 3 & -8 & -8 & (24-\lambda) & -8 \\ 0 & 0 & 2 & 2 & -16 & (22-\lambda) \end{bmatrix} \begin{bmatrix} f_1 \\ f_2 \\ f_3 \\ f_4 \\ f_5 \\ f_6 \end{bmatrix} = \begin{bmatrix} 0 \\ 0 \\ 0 \\ 0 \\ 0 \\ 0 \end{bmatrix}$$

$$(10.25)$$

The vanishing condition for the determinant of the above 6×6 matrix yields a sixth-degree polynomial equation in λ, whose roots (the required eigenvalues) are

$$\lambda_1 = 0.796; \ \lambda_2 = 6.829; \ \lambda_3 = 16.698; \ \lambda_4 = 18.683;$$

$$\lambda_5 = 34.568; \ \lambda_6 = 56.426$$

10.5.2.2 Subspace $S^{(2)}$

From the coefficients of the $\Phi_j^{(2)}$ $(j = 1)$, let

$$W_{14} = W_{15} = W_{16} = W_{17} = -W_{18} = -W_{19} = -W_{20} = -W_{21} = f_1$$

Making the above substitutions into Equation (10.20e) with all other W_i $(i = 1,2,...,13; 22,23,24,25)$ set equal to zero, we obtain

$$[(18-\lambda)][f_1] = [0] \tag{10.26}$$

yielding the solution $\lambda = 18$ when the determinant of the 1×1 matrix is set equal to zero. This is the eigenvalue (natural frequency) corresponding to the one mode of vibration with the symmetry type of subspace $S^{(2)}$. It is noted that for this mode, the four axes of symmetry of the plate $\{x,y,1,2\}$ are all nodal lines experiencing no transverse displacements.

10.5.2.3 Subspace $S^{(3)}$

From the coefficients of the $\Phi_j^{(3)}$ $(j = 1,2,3)$, let

$$W_2 = -W_3 = -W_4 = W_5 = f_1$$

$$W_{10} = -W_{11} = -W_{12} = W_{13} = f_2$$

$$W_{14} = -W_{15} = -W_{16} = W_{17} = W_{18} = W_{19} = -W_{20} = -W_{21} = f_3$$

Making these substitutions into Equations (10.20b), (10.20d) and (10.20e) with all the other W_i $(i = 1,6,7,8,9,22,23,24,25)$ set equal to zero, we obtain three equations:

$$\begin{bmatrix} (17-\lambda) & -8 & 2 \\ -8 & (21-\lambda) & -16 \\ 1 & -8 & (20-\lambda) \end{bmatrix} \begin{bmatrix} f_1 \\ f_2 \\ f_3 \end{bmatrix} = \begin{bmatrix} 0 \\ 0 \\ 0 \end{bmatrix} \tag{10.27}$$

The vanishing condition for the determinant of the above 3×3 matrix yields a third-degree polynomial equation in λ, the roots of which are the three eigenvalues (hence natural frequencies) corresponding to the modes of vibration with the symmetry type of subspace $S^{(3)}$. The results are

$$\lambda_1 = 6.760; \lambda_2 = 16.688; \lambda_3 = 34.551$$

10.5.2.4 Subspace $S^{(4)}$

From the coefficients of the $\Phi_j^{(4)}$ ($j = 1, 2, 3$), let

$$W_6 = -W_7 = -W_8 = W_9 = f_1$$

$$W_{14} = -W_{15} = -W_{16} = W_{17} = -W_{18} = -W_{19} = W_{20} = W_{21} = f_2$$

$$W_{22} = -W_{23} = -W_{24} = W_{25} = f_3$$

with

$$W_1 = W_2 = W_3 = W_4 = W_5 = W_{10} = W_{11} = W_{12} = W_{13} = 0$$

Making these substitutions into Equations (10.20c), (10.20e) and (10.20g), we obtain

$$\begin{bmatrix} (18-\lambda) & -16 & 2 \\ -8 & (22-\lambda) & -8 \\ 2 & -16 & (22-\lambda) \end{bmatrix} \begin{bmatrix} f_1 \\ f_2 \\ f_3 \end{bmatrix} = \begin{bmatrix} 0 \\ 0 \\ 0 \end{bmatrix} \tag{10.28}$$

The vanishing condition for the determinant of the above 3 × 3 matrix yields a third-degree polynomial equation in λ, the roots of which are the three eigenvalues (hence natural frequencies) corresponding to the modes of vibration with the symmetry type of subspace $S^{(4)}$. The results are

$$\lambda_1 = 5.835; \; \lambda_2 = 18.065; \; \lambda_3 = 38.100$$

10.5.2.5 Subspace $S^{(5,2)}$

From the coefficients of the $\Phi_j^{(5,2)}$ ($j = 1, 2, ..., 6$), let

$$W_2 = -W_3 = W_4 = -W_5 = f_1$$

$$W_6 = -W_9 = f_2$$

$$W_{10} = -W_{11} = W_{12} = -W_{13} = f_3$$

$$W_{14} = -W_{17} = -W_{20} = W_{21} = f_4$$

$$W_{15} = -W_{16} = -W_{18} = W_{19} = f_5$$

$$W_{22} = -W_{25} = f_6$$

with

$$W_1 = W_7 = W_8 = W_{23} = W_{24} = 0$$

Making these substitutions into Equations (10.20b) to (10.20g), we obtain

$$
\begin{bmatrix}
(19-\lambda) & -8 & -8 & 3 & -1 & 0 \\
-16 & (20-\lambda) & 4 & -16 & 0 & 2 \\
-8 & 2 & (21-\lambda) & -8 & 8 & 1 \\
3 & -8 & -8 & (23-\lambda) & -1 & -8 \\
-1 & 0 & 8 & -1 & (19-\lambda) & 0 \\
0 & 2 & 2 & -16 & 0 & (22-\lambda)
\end{bmatrix}
\begin{bmatrix}
f_1 \\ f_2 \\ f_3 \\ f_4 \\ f_5 \\ f_6
\end{bmatrix}
=
\begin{bmatrix}
0 \\ 0 \\ 0 \\ 0 \\ 0 \\ 0
\end{bmatrix}
$$

$$(10.29)$$

The vanishing condition for the determinant of this 6×6 matrix yields a sixth-degree polynomial equation in λ, the roots of which are the six eigenvalues (hence natural frequencies) corresponding to the modes of vibration with the symmetry type of subspace $S^{(5,2)}$. The results are

$$\lambda_1 = 2.776;\ \lambda_2 = 11.284;\ \lambda_3 = 12.091;\ \lambda_4 = 23.724;\ \lambda_5 = 27.401;\ \lambda_6 = 46.724$$

As already explained, subspace $S^{(5,1)}$ has exactly the same set of eigenvalues as subspace $S^{(5,2)}$.

10.5.3 Remarks on the structure of symmetry-adapted eigenvalue matrices

Unlike the $n \times n$ eigenvalue matrix of the conventional finite-difference formulation, the $r \times r$ symmetry-adapted eigenvalue matrix for a given subspace $S^{(i)}$ of dimension r is, in general, not symmetric. This is because the number of ϕ_k components making up each basis vector $\Phi_j^{(i)}$ ($j = 1, 2,...,r$) of the subspace are generally different. The number of components making up a basis vector may be thought of as the *weight* of the basis vector.

In the case of the rectangular plate, each subspace $S^{(i)}$ ($i = 1, 2, 3, 4$) is spanned by six basis vectors $\Phi_j^{(i)}$ ($j = 1, 2,..., 6$), all of which have the same number of basis-vector components ϕ_k (i.e. four components each). Thus, all basis vectors of a given subspace have the same weight. That is why all four 6×6 symmetry-adapted eigenvalue matrices for this problem are symmetric.

In the case of the square plate, the various subspaces are spanned by basis vectors with different numbers of components. Take subspace $S^{(3)}$, for example.

This is three-dimensional (i.e. it is spanned by three basis vectors). The first basis vector $\Phi_1^{(3)}$ has four components $\{\phi_2, \phi_3, \phi_4, \phi_5\}$, the second basis vector $\Phi_2^{(3)}$ also has four components $\{\phi_{10}, \phi_{11}, \phi_{12}, \phi_{13}\}$, but the third basis vector $\Phi_3^{(3)}$ has eight components $\{\phi_{14}, \phi_{15}, \phi_{16}, \phi_{17}, \phi_{18}, \phi_{19}, \phi_{20}, \phi_{21}\}$. Let us write the associated eigenvalue matrix in the symbolic form

$$
A = \begin{bmatrix} a_{11} & a_{12} & a_{13} \\ a_{21} & a_{22} & a_{23} \\ a_{31} & a_{32} & a_{33} \end{bmatrix}
\tag{10.30}
$$

where columns $1, 2, 3$ correspond to basis vectors $\Phi_1^{(3)}, \Phi_2^{(3)}, \Phi_3^{(3)}$. If we examine the obtained results for this matrix, we observe that $a_{12} = a_{21}(= -8)$ since $\Phi_1^{(3)}$ and $\Phi_2^{(3)}$ have the same number of basis-vector components (i.e. these basis vectors have the same weight). On the other hand, $a_{13} = 2a_{31}$ and $a_{23} = 2a_{32}$ since basis vector $\Phi_3^{(3)}$ has twice the number of components of either $\Phi_1^{(3)}$ or $\Phi_2^{(3)}$ (i.e. basis vector $\Phi_3^{(3)}$ has double the weight of $\Phi_1^{(3)}$ or $\Phi_2^{(3)}$). This explains why in general symmetry-adapted eigenvalue matrices cannot be expected to be symmetric.

10.6 CONCLUDING REMARKS

In this chapter, we have presented a group-theoretic procedure for the implementation of the finite-difference method in the solution of plate eigenvalue problems. Although the detailed illustrative procedure has been based on rectangular and square plates (which belong to the symmetry groups C_{2v} and C_{4v}, respectively), it is clear that the procedure is quite general, being also applicable to triangular and hexagonal finite-difference mesh patterns that belong to the symmetry groups C_{3v} and C_{6v}, respectively.

Unlike in the conventional approach where we need to write down the finite-difference equations for all the nodes of the mesh and then solve the ensuing (usually large) eigenvalue problem, the group-theoretic procedure requires the writing down of the finite-difference equations for a relatively small set of nodes (the *generator* nodes), which are then transformed into symmetry-adapted equations for the various subspaces (by means of the basis vectors for each subspace).

In this way, the original eigenvalue problem is decomposed into a series of smaller eigenvalue problems corresponding to the various subspaces, which can be solved independently for the eigenvalues. Importantly, the eigenvalues yielded within the various subspaces are in fact the actual eigenvalues associated with the original problem, and no further conversion is required.

REFERENCES

1. S.P. Timoshenko and S. Woinowsky-Krieger. *Theory of Plates and Shells*. McGraw-Hill, New York, 1959.
2. R. Szilard. *Theory and Analysis of Plates: Classical and Numerical Methods*. Prentice-Hall, Englewood Cliffs, NJ, 1974.
3. A. Zingoni. A Group-Theoretic Finite-Difference Formulation for Plate Eigenvalue Problems. *Computers and Structures*, 2012, 112/113, 266–282.

Chapter 11

Finite-element formulations for symmetric elements

11.1 GROUP-THEORETIC FORMULATION FOR FINITE ELEMENTS

In Chapter 10, we saw how ideas of group theory can be incorporated into the finite-difference method to improve the performance of this numerical method. The same ideas can also be applied to another well-established numerical method, namely, the finite-element method (FEM).

In this chapter, we present a group-theoretic formulation for the efficient computation of matrices for symmetric finite elements, and then apply this to the derivation of consistent mass matrices for a variety of elements, namely, the two-node 2 degree-of-freedom (d.o.f.) truss element, the two-node 4 d.o.f. beam element, the four-node 8 d.o.f. rectangular plane–stress element, the four-node 12 d.o.f. rectangular plate-bending element and the eight-node 24 d.o.f. rectangular hexahedral element. Some of these consistent mass matrices were derived in Chapter 5 using the conventional approach. The general approach is similar to that followed by Zlokovic [1], while the examples presented here are drawn from an earlier publication of the author [2].

The key strategy of the present group-theoretic approach is to decompose the general displacement field of the element into smaller fields of symmetry types corresponding to the subspaces of the symmetry group describing the nodal configuration of the element. In this way, the vector space of the original problem is decomposed into independent subspaces, permitting element shape functions of very simple form to be written down for each subspace, and associated element matrices to be computed through the integration of much simpler terms than those associated with the conventional approach.

Overall, even allowing for the cost of decomposition of the problem and final superposition and transformation of subspace quantities, the computational effort expended in arriving at the results is substantially smaller than that associated with the conventional approach. The benefit is particularly significant for elements with a large number of nodes or degrees of freedom.

11.2 COORDINATE SYSTEM, NODE NUMBERING AND POSITIVE DIRECTIONS

The group-theoretic formulation requires a special choice of origin, sequence of node numbering and positive directions for freedoms. Figures 11.1 to 11.3 show the symmetry-adapted conventions for node numbering and positive directions of nodal freedoms for truss, beam, rectangular plane–stress, rectangular plate-bending and rectangular solid elements. The origin of the local coordinate system of the element is chosen at the centre of symmetry O of the configuration, that is, the point through which all axes of rotation pass, or all planes of reflection intersect.

For line elements in the xy plane, oriented along the x axis and with their perpendicular bisector defining the y axis, node 1 must be chosen on the positive side of the perpendicular bisector, that is, on the positive branch of the x axis. For rectangular flat elements, node 1 is chosen in the positive-positive quadrant of the xy Cartesian coordinate system. For rectangular solid elements, node 1 is chosen in the positive-positive-positive octant of the xyz coordinate system.

Consider a finite element having a nodal configuration belonging to the symmetry group G. Let the collection of all the nodal positions of the element that are permuted (or interchanged among themselves) by symmetry operations of G be referred to as a *nodal set*. The concept of a nodal set was used in Chapter 10. An element may have one or more nodal sets. Finite elements having a large number of nodes with configurations belonging to symmetry groups of low order (that is, having a small number of symmetry elements) generally exhibit a multiplicity of nodal sets.

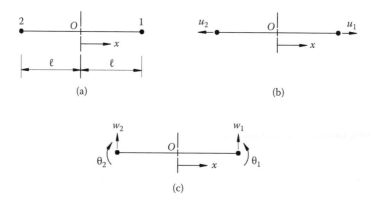

Figure 11.1 Symmetry-adapted conventions for two-node line elements: (a) node numbering, (b) positive directions for nodal freedoms for a two d.o.f. truss element, (c) positive directions for nodal freedoms for a four d.o.f. beam element.

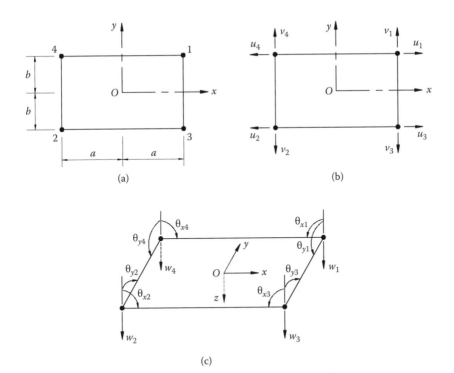

Figure 11.2 Symmetry-adapted conventions for four-node rectangular plane elements: (a) node numbering, (b) positive directions for nodal freedoms for an eight d.o.f. plane–stress element, (c) positive directions for nodal freedoms for a twelve d.o.f. plate-bending element.

The rest of the nodes belonging to the same nodal set as node 1 are numbered consecutively in the order generated by permuting node 1 through each symmetry operation of G, such symmetry operations being performed in the same order as appears across the top of the character table of the symmetry group. Character tables of groups C_{1v}, C_{2v} and C_{4v} were given in Chapter 6. Character tables of other symmetry groups may be seen in references [3, 4].

For symmetry groups C_{1v}, C_{2v} and D_{2h} describing the configurations of straight line elements, plane rectangular elements and solid rectangular elements, respectively, the symmetry elements appearing across the top of the respective character tables are $\{e, C_2\}$, $\{e, C_2, \sigma_x, \sigma_y\}$ and $\{e, C_2^z, C_2^y, C_2^x, i, \sigma_{xy}, \sigma_{xz}, \sigma_{yz}\}$. Here e is the identity element, and as in earlier chapters, C_2 is a rotation through 180°, σ_x is a reflection in the perpendicular plane containing the x axis and σ_y is a reflection in the perpendicular plane containing the y axis. For symmetry group D_{2h} associated with

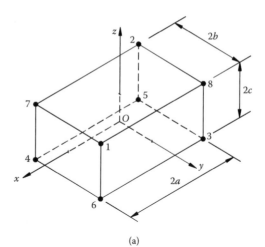

(a)

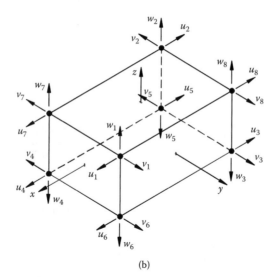

(b)

Figure 11.3 Symmetry-adapted conventions for an eight-node rectangular solid element: (a) node numbering, (b) positive directions for nodal freedoms for a 24 d.o.f. hexahedral element.

the three-dimensional element, C_2^z is a rotation through $180°$ about the z axis, C_2^y is a rotation through $180°$ about the y axis, C_2^x is a rotation through $180°$ about the x axis, i is an inversion about the centre of symmetry O, σ_{xy} is a reflection in the xy plane, σ_{xz} is a reflection in the xz plane and σ_{yz} is a reflection in the yz plane.

After all the nodal positions of the first nodal set have been numbered, say, from 1 up to n_1, another node on the positive branch of the x axis (in the case of line elements), or in the positive-positive quadrant of the xy coordinate system (in the case of rectangular plane elements), or in the positive-positive-positive octant of the xyz coordinate system (in the case of rectangular solid elements), is selected and numbered $n_1 + 1$, and the rest of the nodal positions in this second nodal set are numbered consecutively upwards from $n_1 + 2$ following the same sequence described for the first nodal set. The procedure is repeated for all nodal sets of the element's nodal positions until all n individual nodes of the element have been numbered consecutively from 1 up to n.

Having numbered all the nodes in accordance with the above procedure, the positive directions of the freedoms and loads at node 1 may be chosen arbitrarily, but having fixed these, the positive directions for the rest of the nodal freedoms and nodal loads of the element should be chosen such that the ensuing overall pattern of directions of nodal quantities has the symmetry type of the first irreducible representation of the symmetry group G (refer to Chapter 6). In other words, the pattern of nodal directions must preserve the full symmetry of the pattern of nodes itself. According to this scheme, corresponding nodes of the various nodal sets will have the same positive directions of nodal quantities.

The exception to this rule is the case of nodes that coincide with the symmetry axes or symmetry planes. The positive directions for these cannot, in general, be chosen to conform to the full symmetry of the pattern of nodes; they may be taken in the same directions as for node 1.

11.3 SYMMETRY-ADAPTED NODAL FREEDOMS

Given a finite element with a nodal configuration belonging to the symmetry group G, we can use the idempotents of G to transform the normal freedoms of the problem into sets of symmetry-adapted freedoms spanning the subspaces of the problem. The idempotents $P^{(j)}$ of symmetry groups C_{1v}, C_{2v} and D_{2h}, where $j = \{1,2\}$ for group C_{1v}, $j = \{1,2,3,4\}$ for group C_{2v} and $j = \{1,2, ..., 8\}$ for group D_{2h}, may be written in the following compact forms:

$$\begin{bmatrix} P^{(1)} \\ P^{(2)} \end{bmatrix} = \frac{1}{2}\begin{bmatrix} 1 & 1 \\ 1 & -1 \end{bmatrix}\begin{bmatrix} e \\ \sigma_y \end{bmatrix} \tag{11.1}$$

$$\begin{bmatrix} P^{(1)} \\ P^{(2)} \\ P^{(3)} \\ P^{(4)} \end{bmatrix} = \frac{1}{4}\begin{bmatrix} 1 & 1 & 1 & 1 \\ 1 & 1 & -1 & -1 \\ 1 & -1 & 1 & -1 \\ 1 & -1 & -1 & 1 \end{bmatrix}\begin{bmatrix} e \\ C_2 \\ \sigma_x \\ \sigma_y \end{bmatrix} \tag{11.2}$$

$$
\begin{bmatrix} P^{(1)} \\ P^{(2)} \\ P^{(3)} \\ P^{(4)} \\ P^{(5)} \\ P^{(6)} \\ P^{(7)} \\ P^{(8)} \end{bmatrix} = \frac{1}{8} \begin{bmatrix} 1 & 1 & 1 & 1 & 1 & 1 & 1 & 1 \\ 1 & 1 & -1 & -1 & 1 & 1 & -1 & -1 \\ 1 & -1 & 1 & -1 & 1 & -1 & 1 & -1 \\ 1 & -1 & -1 & 1 & 1 & -1 & -1 & 1 \\ 1 & 1 & 1 & 1 & -1 & -1 & -1 & -1 \\ 1 & 1 & -1 & -1 & -1 & -1 & 1 & 1 \\ 1 & -1 & 1 & -1 & -1 & 1 & -1 & 1 \\ 1 & -1 & -1 & 1 & -1 & 1 & 1 & -1 \end{bmatrix} \begin{bmatrix} e \\ C_2^z \\ C_2^y \\ C_2^x \\ i \\ \sigma_{xy} \\ \sigma_{xz} \\ \sigma_{yz} \end{bmatrix}
\tag{11.3}
$$

The above relations may be written in the general form $\mathbf{P} = \mathbf{T\Omega}$, where $\mathbf{P}$ is the column vector of idempotents, $\mathbf{\Omega}$ is the column vector of symmetry elements of G, and $\mathbf{T}$ is the square matrix linking the two vectors.

If we apply an idempotent $P^{(j)}$ to the nodal-freedom sets $D_{(i)}$ of a finite-element configuration, we obtain *symmetry-adapted* nodal-freedom sets $\Phi^{(j)}$ corresponding to the subspace $S^{(j)}$ of the problem.

For instance, considering a rectangular hexahedral element with an arbitrary number n of nodes, applying the first idempotent $P^{(1)}$ of symmetry group D_{2h} to each of the nodal-freedom sets $D_{(i)}$ ($i = 1,..,n$) of the configuration, we obtain n linear combinations of the $D_{(i)}$ (with up to eight different $D_{(i)}$ per linear combination), from which we can identify a set of k_1 ($k_1 < n$) *independent* linear combinations, namely, $\phi_1^{(1)}, \phi_2^{(1)}, .., \phi_{k_1}^{(1)}$, as the basis vectors or symmetry-adapted nodal-freedom sets for subspace $S^{(1)}$. Similarly, the second idempotent $P^{(2)}$ yields k_2 symmetry-adapted nodal-freedom sets for subspace $S^{(2)}$, namely, $\phi_1^{(2)}, \phi_2^{(2)}, .., \phi_{k_2}^{(2)}$, and so on.

For the rectangular hexahedral element of n nodes, the k_j ($j = 1,2,..,8$) symmetry-adapted nodal-freedom sets for subspace $S^{(j)}$, namely, $\phi_1^{(j)}, \phi_2^{(j)}, .., \phi_{k_j}^{(j)}$, may be collected together as the column vector $\Phi^{(j)}$. Thus, the dimensions of subspaces $S^{(1)}, S^{(2)}, .., S^{(8)}$ are $k_1, k_2, .., k_8$, respectively, and $k_1 + k_2 + .. + k_8 = n$.

The results for symmetry-adapted nodal-freedom sets for the two-node 2 d.o.f. truss element, two-node 4 d.o.f. beam element, four-node 8 d.o.f. rectangular plane–stress element, four-node 12 d.o.f. rectangular plate-bending element and eight-node 24 d.o.f. rectangular hexahedral element are given below as Equations (11.4) to (11.8), respectively.

$$
\begin{bmatrix} \Phi^{(1)} \\ \Phi^{(2)} \end{bmatrix} = \begin{bmatrix} \phi_1^{(1)} \\ \phi_1^{(2)} \end{bmatrix} = \frac{1}{2} \begin{bmatrix} (u_1 + u_2) \\ (u_1 - u_2) \end{bmatrix} = \frac{1}{2} \begin{bmatrix} 1 & 1 \\ 1 & -1 \end{bmatrix} \begin{bmatrix} u_1 \\ u_2 \end{bmatrix}
$$

$$
= \frac{1}{2} \begin{bmatrix} 1 & 1 \\ 1 & -1 \end{bmatrix} \begin{bmatrix} D_{(1)} \\ D_{(2)} \end{bmatrix}
\tag{11.4}
$$

$$
\begin{bmatrix} \Phi^{(1)} \\ \Phi^{(2)} \end{bmatrix} = \begin{bmatrix} \left[\begin{array}{cc} \phi_1^{(1)} & \phi_2^{(1)} \end{array} \right]^{\mathrm{T}} \\ \left[\begin{array}{cc} \phi_1^{(2)} & \phi_2^{(2)} \end{array} \right]^{\mathrm{T}} \end{bmatrix} = \frac{1}{2} \begin{bmatrix} \left[\begin{array}{cc} (w_1 + w_2) & (\theta_1 + \theta_2) \end{array} \right]^{\mathrm{T}} \\ \left[\begin{array}{cc} (w_1 - w_2) & (\theta_1 - \theta_2) \end{array} \right]^{\mathrm{T}} \end{bmatrix}
$$

$$
= \frac{1}{2} \begin{bmatrix} 1 & 1 \\ 1 & -1 \end{bmatrix} \begin{bmatrix} D_{(1)} \\ D_{(2)} \end{bmatrix}
$$

$$(11.5)$$

$$
\begin{bmatrix} \Phi^{(1)} \\ \Phi^{(2)} \\ \Phi^{(3)} \\ \Phi^{(4)} \end{bmatrix} = \begin{bmatrix} \left[\begin{array}{cc} \phi_1^{(1)} & \phi_2^{(1)} \end{array} \right]^{\mathrm{T}} \\ \left[\begin{array}{cc} \phi_1^{(2)} & \phi_2^{(2)} \end{array} \right]^{\mathrm{T}} \\ \left[\begin{array}{cc} \phi_1^{(3)} & \phi_2^{(3)} \end{array} \right]^{\mathrm{T}} \\ \left[\begin{array}{cc} \phi_1^{(4)} & \phi_2^{(4)} \end{array} \right]^{\mathrm{T}} \end{bmatrix} = \frac{1}{4} \begin{bmatrix} \left[\begin{array}{cc} (u_1 + u_2 + u_3 + u_4) & (v_1 + v_2 + v_3 + v_4) \end{array} \right]^{\mathrm{T}} \\ \left[\begin{array}{cc} (u_1 + u_2 - u_3 - u_4) & (v_1 + v_2 - v_3 - v_4) \end{array} \right]^{\mathrm{T}} \\ \left[\begin{array}{cc} (u_1 - u_2 + u_3 - u_4) & (v_1 - v_2 + v_3 - v_4) \end{array} \right]^{\mathrm{T}} \\ \left[\begin{array}{cc} (u_1 - u_2 - u_3 + u_4) & (v_1 - v_2 - v_3 + v_4) \end{array} \right]^{\mathrm{T}} \end{bmatrix}
$$

$$
= \frac{1}{4} \begin{bmatrix} 1 & 1 & 1 & 1 \\ 1 & 1 & -1 & -1 \\ 1 & -1 & 1 & -1 \\ 1 & -1 & -1 & 1 \end{bmatrix} \begin{bmatrix} D_{(1)} \\ D_{(2)} \\ D_{(3)} \\ D_{(4)} \end{bmatrix}
$$

$$(11.6)$$

$$
\begin{bmatrix} \Phi^{(1)} \\ \Phi^{(2)} \\ \Phi^{(3)} \\ \Phi^{(4)} \end{bmatrix} = \begin{bmatrix} \left[\begin{array}{ccc} \phi_1^{(1)} & \phi_2^{(1)} & \phi_3^{(1)} \end{array} \right]^{\mathrm{T}} \\ \left[\begin{array}{ccc} \phi_1^{(2)} & \phi_2^{(2)} & \phi_3^{(2)} \end{array} \right]^{\mathrm{T}} \\ \left[\begin{array}{ccc} \phi_1^{(3)} & \phi_2^{(3)} & \phi_3^{(3)} \end{array} \right]^{\mathrm{T}} \\ \left[\begin{array}{ccc} \phi_1^{(4)} & \phi_2^{(4)} & \phi_3^{(4)} \end{array} \right]^{\mathrm{T}} \end{bmatrix}
$$

$$
= \frac{1}{4} \begin{bmatrix} \left[\begin{array}{ccc} (w_1 + w_2 + w_3 + w_4) & (\theta_{x1} + \theta_{x2} + \theta_{x3} + \theta_{x4}) & (\theta_{y1} + \theta_{y2} + \theta_{y3} + \theta_{y4}) \end{array} \right]^{\mathrm{T}} \\ \left[\begin{array}{ccc} (w_1 + w_2 - w_3 - w_4) & (\theta_{x1} + \theta_{x2} - \theta_{x3} - \theta_{x4}) & (\theta_{y1} + \theta_{y2} - \theta_{y3} - \theta_{y4}) \end{array} \right]^{\mathrm{T}} \\ \left[\begin{array}{ccc} (w_1 - w_2 + w_3 - w_4) & (\theta_{x1} - \theta_{x2} + \theta_{x3} - \theta_{x4}) & (\theta_{y1} - \theta_{y2} + \theta_{y3} - \theta_{y4}) \end{array} \right]^{\mathrm{T}} \\ \left[\begin{array}{ccc} (w_1 - w_2 - w_3 + w_4) & (\theta_{x1} - \theta_{x2} - \theta_{x3} + \theta_{x4}) & (\theta_{y1} - \theta_{y2} - \theta_{y3} + \theta_{y4}) \end{array} \right]^{\mathrm{T}} \end{bmatrix}
$$

$$
= \frac{1}{4} \begin{bmatrix} 1 & 1 & 1 & 1 \\ 1 & 1 & -1 & -1 \\ 1 & -1 & 1 & -1 \\ 1 & -1 & -1 & 1 \end{bmatrix} \begin{bmatrix} D_{(1)} \\ D_{(2)} \\ D_{(3)} \\ D_{(4)} \end{bmatrix}
$$

$$(11.7)$$

$$
\begin{bmatrix} \Phi^{(1)} \\ \Phi^{(2)} \\ \Phi^{(3)} \\ \Phi^{(4)} \\ \Phi^{(5)} \\ \Phi^{(6)} \\ \Phi^{(7)} \\ \Phi^{(8)} \end{bmatrix}
=
\begin{bmatrix}
\left[\ \phi_1^{(1)}\ \phi_2^{(1)}\ \phi_3^{(1)}\ \right]^{\mathrm T} \\
\left[\ \phi_1^{(2)}\ \phi_2^{(2)}\ \phi_3^{(2)}\ \right]^{\mathrm T} \\
\left[\ \phi_1^{(3)}\ \phi_2^{(3)}\ \phi_3^{(3)}\ \right]^{\mathrm T} \\
\left[\ \phi_1^{(4)}\ \phi_2^{(4)}\ \phi_3^{(4)}\ \right]^{\mathrm T} \\
\left[\ \phi_1^{(5)}\ \phi_2^{(5)}\ \phi_3^{(5)}\ \right]^{\mathrm T} \\
\left[\ \phi_1^{(6)}\ \phi_2^{(6)}\ \phi_3^{(6)}\ \right]^{\mathrm T} \\
\left[\ \phi_1^{(7)}\ \phi_2^{(7)}\ \phi_3^{(7)}\ \right]^{\mathrm T} \\
\left[\ \phi_1^{(8)}\ \phi_2^{(8)}\ \phi_3^{(8)}\ \right]^{\mathrm T}
\end{bmatrix}
\tag{11.8a}
$$

$$
= \frac{1}{8}
\begin{bmatrix}
1 & 1 & 1 & 1 & 1 & 1 & 1 & 1 \\
1 & 1 & -1 & -1 & 1 & 1 & -1 & -1 \\
1 & -1 & 1 & -1 & 1 & -1 & 1 & -1 \\
1 & -1 & -1 & 1 & 1 & -1 & -1 & 1 \\
1 & 1 & 1 & 1 & -1 & -1 & -1 & -1 \\
1 & 1 & -1 & -1 & -1 & -1 & 1 & 1 \\
1 & -1 & 1 & -1 & -1 & 1 & -1 & 1 \\
1 & -1 & -1 & 1 & -1 & 1 & 1 & -1
\end{bmatrix}
\begin{bmatrix} D_{(1)} \\ D_{(2)} \\ D_{(3)} \\ D_{(4)} \\ D_{(5)} \\ D_{(6)} \\ D_{(7)} \\ D_{(8)} \end{bmatrix}
$$

where

$$
\begin{bmatrix} \phi_1^{(1)} \\ \phi_1^{(2)} \\ \phi_1^{(3)} \\ \phi_1^{(4)} \\ \phi_1^{(5)} \\ \phi_1^{(6)} \\ \phi_1^{(7)} \\ \phi_1^{(8)} \end{bmatrix}
= \frac{1}{8}
\begin{bmatrix}
\left(u_1 + u_2 + u_3 + u_4 + u_5 + u_6 + u_7 + u_8\right) \\
\left(u_1 + u_2 - u_3 - u_4 + u_5 + u_6 - u_7 - u_8\right) \\
\left(u_1 - u_2 + u_3 - u_4 + u_5 - u_6 + u_7 - u_8\right) \\
\left(u_1 - u_2 - u_3 + u_4 + u_5 - u_6 - u_7 + u_8\right) \\
\left(u_1 + u_2 + u_3 + u_4 - u_5 - u_6 - u_7 - u_8\right) \\
\left(u_1 + u_2 - u_3 - u_4 - u_5 - u_6 + u_7 + u_8\right) \\
\left(u_1 - u_2 + u_3 - u_4 - u_5 + u_6 - u_7 + u_8\right) \\
\left(u_1 - u_2 - u_3 + u_4 - u_5 + u_6 + u_7 - u_8\right)
\end{bmatrix}
\tag{11.8b}
$$

and exactly similar relations apply for the $\phi_2^{(j)}$ (combinations of $v_1, v_2,...,v_8$) and $\phi_3^{(j)}$ (combinations of $w_1, w_2, ..., w_8$), where $j = 1,..., 8$. All five of the above relations (Equations (11.4) to (11.8)) are of the form $\Phi = TD$, where Φ is the column vector of symmetry-adapted nodal freedoms, D is the column vector of conventional nodal freedoms, and T is as previously defined.

II.4 DISPLACEMENT FIELD DECOMPOSITION

We achieve this by allocating terms of the displacement polynomial associated with the finite element in question, to the subspaces of the symmetry group of its configuration, in accordance with the symmetries of the displacement fields associated with the terms. The assumed polynomials for the displacement components $\{u, v\}$ of rectangular plane–stress elements and $\{u, v, w\}$ of rectangular solid elements are usually of the same form, so we need only consider one displacement component, say u, and allocate the terms of the displacement polynomial to the subspaces of the respective symmetry groups.

The following polynomials are usually adopted to represent the displacement fields of the two-node 2 d.o.f. truss element (Equation (11.9)), the two-node 4 d.o.f. beam element (Equations (11.10)), the four-node 8 d.o.f. rectangular plane–stress element (Equation (11.11)), the four-node 12 d.o.f. rectangular plate-bending element (Equations (11.12)) and the eight-node 24 d.o.f. rectangular hexahedral element (Equation (11.13)):

$$u(x) = \alpha_1 + \alpha_2 x \tag{11.9}$$

$$w(x) = \alpha_1 + \alpha_2 x + \alpha_3 x^2 + \alpha_4 x^3 \tag{11.10a}$$

$$\theta(x) = \frac{dw(x)}{dx} = \alpha_2 + 2\alpha_3 x + 3\alpha_4 x^2 \tag{11.10b}$$

$$u(x, y) = \alpha_1 + \alpha_2 x + \alpha_3 y + \alpha_4 xy \tag{11.11}$$

$$\begin{aligned} w = \alpha_1 + \alpha_2 x + \alpha_3 y + \alpha_4 x^2 + \alpha_5 xy + \alpha_6 y^2 + \alpha_7 x^3 + \alpha_8 x^2 y \\ + \alpha_9 xy^2 + \alpha_{10} y^3 + \alpha_{11} x^3 y + \alpha_{12} xy^3 \end{aligned} \tag{11.12a}$$

$$\begin{aligned} \theta_x = \frac{\partial w}{\partial x} = \alpha_2 + 2x\alpha_4 + \alpha_5 y + 3x^2 \alpha_7 + 2xy\alpha_8 + \alpha_9 y^2 \\ + 3x^2 y\alpha_{11} + \alpha_{12} y^3 \end{aligned} \tag{11.12b}$$

$$\theta_y = \frac{\partial w}{\partial y} = \alpha_3 + \alpha_5 x + 2y\alpha_6 + \alpha_8 x^2 + 2yx\alpha_9 + 3y^2\alpha_{10} + \alpha_{11}x^3 + 3y^2 x\alpha_{12}$$

(11.12c)

$$u(x, y, z) = \alpha_1 + \alpha_2 x + \alpha_3 y + \alpha_4 z + \alpha_5 xy + \alpha_6 yz + \alpha_7 xz + \alpha_8 xyz \quad (11.13)$$

To visualize which terms of a polynomial belong to what subspace, we may plot the implied displacements (positive or negative) at the two ends or the four corners of the element, and then observe the symmetry of the ensuing pattern of nodal displacements. Such plots for the truss element (Equation (11.9)), the rectangular plane–stress element (Equation (11.11)) and the rectangular solid element (Equation (11.13)) are depicted in Figures 11.4 to 11.6.

In this way, terms of the displacement polynomials in Equation (11.9) for the truss element and Equations (11.10) for the beam element have been assigned to the two subspaces $S^{(1)}$ and $S^{(2)}$ of the symmetry group C_{1v}, and the respective results for these two elements are given in Equations (11.14) and (11.15). Similarly, terms of the displacement polynomials in Equation (11.11) for the rectangular plane–stress element and Equations (11.12) for the rectangular plate-bending element have been allocated to the four subspaces $S^{(1)}$, $S^{(2)}$, $S^{(3)}$ and $S^{(4)}$ of the symmetry group C_{2v}, and the respective results for these two elements are given in Equations (11.16) and (11.17).

(a)   (b)

Figure 11.4 Symmetries of the displacement terms of the two-node two d.o.f. truss element (Equation (11.9)): (a) $u = \alpha_1$: subspace $S^{(2)}$, (b) $u = \alpha_2 x$: subspace $S^{(1)}$.

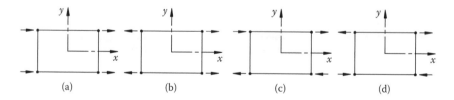

(a)   (b)   (c)   (d)

Figure 11.5 Symmetries of the displacement terms of the four-node eight d.o.f. rectangular plane–stress element (Equation (11.11)): (a) $u = \alpha_1$: subspace $S^{(3)}$, (b) $u = \alpha_2 x$: subspace $S^{(1)}$, (c) $u = \alpha_3 y$: subspace $S^{(2)}$, (d) $u = \alpha_4 xy$: subspace $S^{(4)}$.

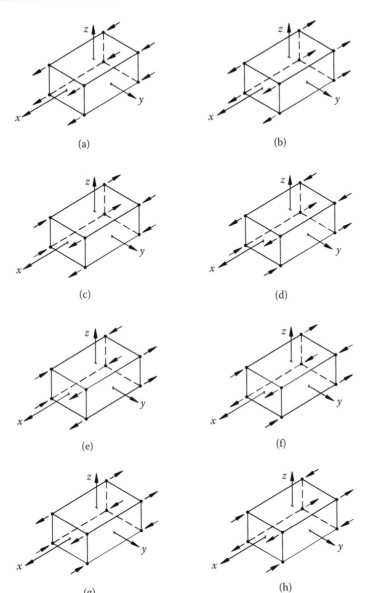

(a)

(b)

(c)

(d)

(e)

(f)

(g)

(h)

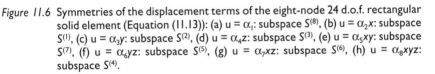

Figure 11.6 Symmetries of the displacement terms of the eight-node 24 d.o.f. rectangular solid element (Equation (11.13)): (a) $u = \alpha_1$: subspace $S^{(8)}$, (b) $u = \alpha_2 x$: subspace $S^{(1)}$, (c) $u = \alpha_3 y$: subspace $S^{(2)}$, (d) $u = \alpha_4 z$: subspace $S^{(3)}$, (e) $u = \alpha_5 xy$: subspace $S^{(7)}$, (f) $u = \alpha_6 yz$: subspace $S^{(5)}$, (g) $u = \alpha_7 xz$: subspace $S^{(6)}$, (h) $u = \alpha_8 xyz$: subspace $S^{(4)}$.

Finally, terms of the displacement polynomial in Equation (11.13) for the rectangular hexahedral element have been allocated to the eight subspaces of the symmetry group D_{2h}, and the results for this element are given in Equation (11.18).

$$
\begin{bmatrix} u^{(1)}(x) \\ u^{(2)}(x) \end{bmatrix} = \begin{bmatrix} x & 0 \\ 0 & 1 \end{bmatrix} \begin{bmatrix} \alpha_2 \\ \alpha_1 \end{bmatrix}
\tag{11.14}
$$

$$
\begin{bmatrix} w^{(1)}(x) \\ \theta^{(1)}(x) \\ w^{(2)}(x) \\ \theta^{(2)}(x) \end{bmatrix} = \begin{bmatrix} 1 & x^2 & & \\ 0 & 2x & & \\ & & x & x^3 \\ & & 1 & 3x^2 \end{bmatrix} \begin{bmatrix} \alpha_1 \\ \alpha_3 \\ \alpha_2 \\ \alpha_4 \end{bmatrix}
\tag{11.15}
$$

$$
\begin{bmatrix} u^{(1)}(x,y) \\ u^{(2)}(x,y) \\ u^{(3)}(x,y) \\ u^{(4)}(x,y) \end{bmatrix} = \begin{bmatrix} x & & & \\ & y & & \\ & & 1 & \\ & & & xy \end{bmatrix} \begin{bmatrix} \alpha_2 \\ \alpha_3 \\ \alpha_1 \\ \alpha_4 \end{bmatrix}
\tag{11.16}
$$

$$
\begin{bmatrix} w^{(1)}(x,y) \\ \theta_x^{(1)}(x,y) \\ \theta_y^{(1)}(x,y) \\ w^{(2)}(x,y) \\ \theta_x^{(2)}(x,y) \\ \theta_y^{(2)}(x,y) \\ w^{(3)}(x,y) \\ \theta_x^{(3)}(x,y) \\ \theta_y^{(3)}(x,y) \\ w^{(4)}(x,y) \\ \theta_x^{(4)}(x,y) \\ \theta_y^{(4)}(x,y) \end{bmatrix} = \begin{bmatrix} \begin{matrix}1 & x^2 & y^2 \\ 0 & 2x & 0 \\ 0 & 0 & 2y\end{matrix} & & & \\ & \begin{matrix}xy & x^3y & xy^3 \\ y & 3x^2y & y^3 \\ x & x^3 & 3xy^2\end{matrix} & & \\ & & \begin{matrix}x & x^3 & xy^2 \\ 1 & 3x^2 & y^2 \\ 0 & 0 & 2xy\end{matrix} & \\ & & & \begin{matrix}y & x^2y & y^3 \\ 0 & 2xy & 0 \\ 1 & x^2 & 3y^2\end{matrix} \end{bmatrix} \begin{bmatrix} \alpha_1 \\ \alpha_4 \\ \alpha_6 \\ \alpha_5 \\ \alpha_{11} \\ \alpha_{12} \\ \alpha_2 \\ \alpha_7 \\ \alpha_9 \\ \alpha_3 \\ \alpha_8 \\ \alpha_{10} \end{bmatrix}
$$

$$
\tag{11.17}
$$

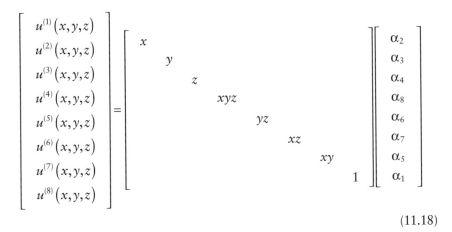

$$(11.18)$$

11.5 SUBSPACE SHAPE FUNCTIONS

If we substitute the nodal coordinates of the first nodes of the nodal sets of the element into the decomposed polynomial(s) for subspace $S^{(i)}$ (as given by Equations (11.14) to (11.18) for the element in question), we automatically obtain the sets $\Phi^{(i)}$ of the symmetry-adapted nodal displacements for subspace $S^{(i)}$, in terms of the polynomial coefficients α_1, α_2, etc.

Rearranging, we obtain the polynomial coefficients (α_1, α_2, etc.) in terms of the symmetry-adapted nodal displacements $\Phi^{(i)}$. Substituting the results into the expressions for the subspace displacement fields (Equations (11.14) to (11.18) as appropriate), we obtain expressions for the symmetry-adapted internal displacements $U^{(i)}$ (which are the subspace displacement fields) in terms of the symmetry-adapted nodal displacements $\Phi^{(i)}$. These final expressions are of the form $U^{(i)} = N^{(i)}\Phi^{(i)}$. By analogy with the conventional finite-element formulation, $N^{(i)}$ is the matrix of *symmetry-adapted shape functions* for subspace $S^{(i)}$.

The results for the symmetry-adapted shape functions thus obtained for the relevant subspaces of the problem, for (1) the two-node 2 d.o.f. truss element, (2) the two-node 4 d.o.f. beam element, (3) the four-node 8 d.o.f. rectangular plane–stress element, (4) the four-node 12 d.o.f. rectangular plate-bending element and (5) the eight-node 24 d.o.f. rectangular hexahedral element, are summarized below in Equations (11.19) to (11.23), respectively.

$$N^{(1)} = \frac{x}{l}; \quad N^{(2)} = 1 \qquad (11.19a, b)$$

$$N_1^{(1)} = 1; \quad N_2^{(1)} = -\frac{l}{2} + \frac{x^2}{2l}$$

(11.20a)

$$N_1^{(2)} = \frac{3x}{2l} - \frac{x^3}{2l^3}; \quad N_2^{(2)} = -\frac{x}{2} + \frac{x^3}{2l^2}$$

(11.20b)

$$N^{(1)} = \frac{x}{a}; \quad N^{(2)} = \frac{y}{b}; \quad N^{(3)} = 1; \quad N^{(4)} = \frac{xy}{ab}$$

(11.21a–d)

$$N_1^{(1)} = 1; \quad N_2^{(1)} = -\frac{a}{2} + \frac{x^2}{2a}; \quad N_3^{(1)} = -\frac{b}{2} + \frac{y^2}{2b}$$

(11.22a)

$$N_1^{(2)} = \frac{2xy}{ab} - \frac{x^3y}{2a^3b} - \frac{xy^3}{2ab^3}; \quad N_2^{(2)} = -\frac{xy}{2b} + \frac{x^3y}{2a^2b}; \quad N_3^{(2)} = -\frac{xy}{2a} + \frac{xy^3}{2ab^2}$$

(11.22b)

$$N_1^{(3)} = \frac{3x}{2a} - \frac{x^3}{2a^3}; \quad N_2^{(3)} = -\frac{x}{2} + \frac{x^3}{2a^2}; \quad N_3^{(3)} = -\frac{bx}{2a} + \frac{xy^2}{2ab}$$

(11.22c)

$$N_1^{(4)} = \frac{3y}{2b} - \frac{y^3}{2b^3}; \quad N_2^{(4)} = -\frac{ya}{2b} + \frac{x^2y}{2ab}; \quad N_3^{(4)} = -\frac{y}{2} + \frac{y^3}{2b^2}$$

(11.22d)

$$N^{(1)} = \frac{x}{a}; \quad N^{(2)} = \frac{y}{b}; \quad N^{(3)} = \frac{z}{c}; \quad N^{(4)} = \frac{xyz}{abc}$$

(11.23a–d)

$$N^{(5)} = \frac{yz}{bc}; \quad N^{(6)} = \frac{xz}{ac}; \quad N^{(7)} = \frac{xy}{ab}; \quad N^{(8)} = 1$$

(11.23e–h)

We observe that the subspace shape functions are much simpler sets of expressions in comparison with conventional shape functions for these elements.

11.6 SUBSPACE ELEMENT MATRICES

Symmetry-adapted element matrices in the subspaces of a problem are obtained by integrating the symmetry-adapted shape functions over a representative domain of the element. For line elements, this integration is done over only the positive half of the element; for rectangular plane elements, we integrate over only the positive-positive quadrant of the element, and for rectangular solid elements, we integrate over only the positive-positive-positive octant.

In dynamic finite-element modelling (refer to Chapter 5), the element stiffness and consistent mass matrices are of interest. For a given element, the symmetry-adapted stiffness matrix $K^{(i)}$ and consistent mass matrix $M^{(i)}$ for subspace $S^{(i)}$ are obtained from the relationships

$$K^{(i)} = \int_V \left[B^{(i)} \right]^T [D_m] \left[B^{(i)} \right] dV; \quad M^{(i)} = \int_V \rho \left[N^{(i)} \right]^T \left[N^{(i)} \right] dV \quad (11.24)$$

where $B^{(i)}$ and $N^{(i)}$ are the subspace strain displacement and shape function matrices, respectively, D_m is the matrix of material stiffnesses and ρ is the material mass density. These are similar to the conventional relationships of Chapter 5, except that here they pertain to subspaces, with the integration being performed over a reduced domain, as already pointed out.

Noting that the truss and beam elements (Figure 11.1) are of length $2l$, the rectangular plane–stress and plate-bending elements (Figure 11.2) are of dimensions $2a \times 2b$, and the rectangular hexahedral element (Figure 11.3) is of dimensions $2a \times 2b \times 2c$, we perform the integrations for all the subspaces of a given element, over the relevant reduced domains.

For (1) the two-node 2 d.o.f. truss element, (2) the two-node 4 d.o.f. beam element, (3) the four-node 8 d.o.f. rectangular plane–stress element, (4) the four-node 12 d.o.f. rectangular plate-bending element and (5) the eight-node 24 d.o.f. rectangular hexahedral element, the results for the subspace consistent mass matrices (relevant in a dynamic analysis) are given by equation sets (11.25) to (11.29), in that order, as follows:

$$M^{(1)} = \rho A \int_0^l N^{(1)} N^{(1)} \, dx = \frac{\rho A l}{3} \quad (11.25a)$$

$$M^{(2)} = \rho A \int_0^l N^{(2)} N^{(2)} \, dx = \rho A l \quad (11.25b)$$

$$M^{(1)} = \rho A \int_0^l \left[N^{(1)} \right]^T \left[N^{(1)} \right] dx = \frac{\rho A l}{105} \begin{bmatrix} 105 & -35l \\ -35l & 14l^2 \end{bmatrix} \quad (11.26a)$$

$$M^{(2)} = \rho A \int_0^l \left[N^{(2)} \right]^T \left[N^{(2)} \right] dx = \frac{\rho A l}{105} \begin{bmatrix} 51 & -9l \\ -9l & 2l^2 \end{bmatrix} \quad (11.26b)$$

$$M^{(1)} = \rho t \int_0^b \int_0^a N^{(1)}N^{(1)} \, dx \, dy = \frac{\rho tab}{3} \tag{11.27a}$$

$$M^{(2)} = \rho t \int_0^b \int_0^a N^{(2)}N^{(2)} \, dx \, dy = \frac{\rho\, tab}{3} \tag{11.27b}$$

$$M^{(3)} = \rho t \int_0^b \int_0^a N^{(3)}N^{(3)} \, dx \, dy = \rho tab \tag{11.27c}$$

$$M^{(4)} = \rho t \int_0^b \int_0^a N^{(4)}N^{(4)} \, dx \, dy = \frac{\rho\, tab}{9} \tag{11.27d}$$

$$M^{(1)} = \rho t \int_0^b \int_0^a \left[N^{(1)}\right]^T \left[N^{(1)}\right] = \frac{\rho\, tab}{135} \begin{bmatrix} 135 & -45a & -45b \\ -45a & 18a^2 & 15ab \\ -45b & 15ab & 18b^2 \end{bmatrix} \tag{11.28a}$$

$$M^{(2)} = \rho t \int_0^b \int_0^a \left[N^{(2)}\right]^T \left[N^{(2)}\right] = \frac{\rho\, t a b}{1575} \begin{bmatrix} 349 & -52a & -52b \\ -52a & 10a^2 & 7ab \\ -52b & 7ab & 10b^2 \end{bmatrix} \tag{11.28b}$$

$$M^{(3)} = \rho t \int_0^b \int_0^a \left[N^{(3)}\right]^T \left[N^{(3)}\right] = \frac{\rho\, tab}{315} \begin{bmatrix} 153 & -27a & -42b \\ -27a & 6a^2 & 7ab \\ -42b & 7ab & 14b^2 \end{bmatrix} \tag{11.28c}$$

$$M^{(4)} = \rho t \int_0^b \int_0^a \left[N^{(4)}\right]^T \left[N^{(4)}\right] = \frac{\rho\, tab}{315} \begin{bmatrix} 153 & -42a & -27b \\ -42a & 14a^2 & 7ab \\ -27b & 7ab & 6b^2 \end{bmatrix} \tag{11.28d}$$

$$M^{(1)} = \rho \int_0^c \int_0^b \int_0^a N^{(1)}N^{(1)} \, dx \, dy \, dz = \frac{\rho \, abc}{3} \qquad (11.29a)$$

$$M^{(2)} = \rho \int_0^c \int_0^b \int_0^a N^{(2)}N^{(2)} \, dx \, dy \, dz = \frac{\rho \, abc}{3} \qquad (11.29b)$$

$$M^{(3)} = \rho \int_0^c \int_0^b \int_0^a N^{(3)}N^{(3)} \, dx \, dy \, dz = \frac{\rho \, abc}{3} \qquad (11.29c)$$

$$M^{(4)} = \rho \int_0^c \int_0^b \int_0^a N^{(4)}N^{(4)} \, dx \, dy \, dz = \frac{\rho \, abc}{27} \qquad (11.29d)$$

$$M^{(5)} = \rho \int_0^c \int_0^b \int_0^a N^{(5)}N^{(5)} \, dx \, dy \, dz = \frac{\rho \, abc}{9} \qquad (11.29e)$$

$$M^{(6)} = \rho \int_0^c \int_0^b \int_0^a N^{(6)}N^{(6)} \, dx \, dy \, dz = \frac{\rho \, abc}{9} \qquad (11.29f)$$

$$M^{(7)} = \rho \int_0^c \int_0^b \int_0^a N^{(7)}N^{(7)} \, dx \, dy \, dz = \frac{\rho \, abc}{9} \qquad (11.29g)$$

$$M^{(8)} = \rho \int_0^c \int_0^b \int_0^a N^{(8)}N^{(8)} \, dx \, dy \, dz = \rho \, abc \qquad (11.29h)$$

11.7 FINAL ELEMENT MATRICES

We may readily convert the results of the preceding section into conventional form by superposition of the subspace matrices, followed by simple coordinate transformations. For each element, we begin by writing down the permutation table for all the nodes of the element, which is a table showing how the individual nodes are permuted under the operations of the symmetry group of the configuration in question.

The permutation tables for the two-node truss and beam elements (Figure 11.1(a)), the four-node rectangular plane–stress and plate-bending elements (Figure 11.2(a)), and the eight-node rectangular solid element (Figure 11.3(a)) are shown in Tables 11.1 to 11.3. By reference to these

Table 11.1 Permutation table for the two-node elements

Node	e	σ_y
1	1	2
2	2	1

Table 11.2 Permutation table for the four-node rectangular elements

Node	e	C_2	σ_x	σ_y
1	1	2	3	4
2	2	1	4	3
3	3	4	1	2
4	4	3	2	1

Table 11.3 Permutation table for the eight-node hexahedral elements

Node	e	C_2^z	C_2^y	C_2^x	i	σ_{xy}	σ_{xz}	σ_{yz}
1	1	2	3	4	5	6	7	8
2	2	1	4	3	6	5	8	7
3	3	4	1	2	7	8	5	6
4	4	3	2	1	8	7	6	5
5	5	6	7	8	1	2	3	4
6	6	5	8	7	2	1	4	3
7	7	8	5	6	3	4	1	2
8	8	7	6	5	4	3	2	1

tables, the element consistent mass matrices in conventional form may then be written in the forms

$$M = \begin{bmatrix} m_1 & m_2 \\ m_2 & m_1 \end{bmatrix} \tag{11.30}$$

$$M = \begin{bmatrix} m_1 & m_2 & m_3 & m_4 \\ m_2 & m_1 & m_4 & m_3 \\ m_3 & m_4 & m_1 & m_2 \\ m_4 & m_3 & m_2 & m_1 \end{bmatrix} \tag{11.31}$$

$$\mathbf{M} = \begin{bmatrix} m_1 & m_2 & m_3 & m_4 & m_5 & m_6 & m_7 & m_8 \\ m_2 & m_1 & m_4 & m_3 & m_6 & m_5 & m_8 & m_7 \\ m_3 & m_4 & m_1 & m_2 & m_7 & m_8 & m_5 & m_6 \\ m_4 & m_3 & m_2 & m_1 & m_8 & m_7 & m_6 & m_5 \\ m_5 & m_6 & m_7 & m_8 & m_1 & m_2 & m_3 & m_4 \\ m_6 & m_5 & m_8 & m_7 & m_2 & m_1 & m_4 & m_3 \\ m_7 & m_8 & m_5 & m_6 & m_3 & m_4 & m_1 & m_2 \\ m_8 & m_7 & m_6 & m_5 & m_4 & m_3 & m_2 & m_1 \end{bmatrix} \tag{11.32}$$

for (1) the two-node truss and beam elements (Equation (11.30)), (2) the four-node rectangular plane–stress and plate-bending elements (Equation (11.31)) and (3) the eight-node rectangular solid element (Equation (11.32)), the subscripts being given by the entries of the permutation tables.

The elements m_i of the above matrices are linear combinations of the respective subspace matrices $\mathbf{M}^{(j)}$ (refer to Equations (11.25) to (11.29)), with the signs of the terms (+ or –) being given by row i of the corresponding matrices $\mathbf{T}$ (refer to Equations (11.1) to (11.3)).

The results for (1) the two-node 2 d.o.f. truss element, (2) the two-node 4 d.o.f. beam element, (3) the four-node 8 d.o.f. rectangular plane–stress element, (4) the four-node 12 d.o.f. rectangular plate-bending element and (5) the eight-node 24 d.o.f. rectangular hexahedral element, are given by Equations (11.33) to (11.37), respectively, as follows:

$$m_1 = \frac{1}{2}\left(\mathbf{M}^{(1)} + \mathbf{M}^{(2)}\right) = \frac{2\rho Al}{3} \tag{11.33a}$$

$$m_2 = \frac{1}{2}\left(\mathbf{M}^{(1)} - \mathbf{M}^{(2)}\right) = -\frac{\rho Al}{3} \tag{11.33b}$$

$$m_1 = \frac{1}{2}\left(\mathbf{M}^{(1)} + \mathbf{M}^{(2)}\right) = \frac{\rho Al}{210}\begin{bmatrix} 156 & -22l \\ -22l & 4l^2 \end{bmatrix} \tag{11.34a}$$

$$m_2 = \frac{1}{2}\left(\mathbf{M}^{(1)} - \mathbf{M}^{(2)}\right) = \frac{\rho Al}{210}\begin{bmatrix} 54 & -13l \\ -13l & 3l^2 \end{bmatrix} \tag{11.34b}$$

$$m_1 = \frac{1}{4}\left(\mathbf{M}^{(1)} + \mathbf{M}^{(2)} + \mathbf{M}^{(3)} + \mathbf{M}^{(4)}\right) = \frac{4\rho\, tab}{9} \tag{11.35a}$$

$$m_2 = \frac{1}{4}\left(M^{(1)} + M^{(2)} - M^{(3)} - M^{(4)}\right) = -\frac{\rho \, tab}{9} \tag{11.35b}$$

$$m_3 = \frac{1}{4}\left(M^{(1)} - M^{(2)} + M^{(3)} - M^{(4)}\right) = \frac{2\rho \, tab}{9} \tag{11.35c}$$

$$m_4 = \frac{1}{4}\left(M^{(1)} - M^{(2)} - M^{(3)} + M^{(4)}\right) = -\frac{2\rho \, tab}{9} \tag{11.35d}$$

$$m_1 = \frac{1}{4}\left(M^{(1)} + M^{(2)} + M^{(3)} + M^{(4)}\right) = \frac{\rho \, tab}{3150}\begin{bmatrix} 1727 & -461a & -461b \\ -461a & 160a^2 & 126ab \\ -461b & 126ab & 160b^2 \end{bmatrix}$$

$$\tag{11.36a}$$

$$m_2 = \frac{1}{4}\left(M^{(1)} + M^{(2)} - M^{(3)} - M^{(4)}\right) = \frac{\rho \, tab}{3150}\begin{bmatrix} 197 & -116a & -116b \\ -116a & 60a^2 & 56ab \\ -116b & 56ab & 60b^2 \end{bmatrix}$$

$$\tag{11.36b}$$

$$m_3 = \frac{1}{4}\left(M^{(1)} - M^{(2)} + M^{(3)} - M^{(4)}\right) = \frac{\rho \, tab}{3150}\begin{bmatrix} 613 & -199a & -274b \\ -199a & 80a^2 & 84ab \\ -274b & 84ab & 120b^2 \end{bmatrix}$$

$$\tag{11.36c}$$

$$m_4 = \frac{1}{4}\left(M^{(1)} - M^{(2)} - M^{(3)} + M^{(4)}\right) = \frac{\rho \, tab}{3150}\begin{bmatrix} 613 & -274a & -199b \\ -274a & 120a^2 & 84ab \\ -199b & 84ab & 80b^2 \end{bmatrix}$$

$$\tag{11.36d}$$

$$m_1 = \frac{1}{8}\left(M^{(1)} + M^{(2)} + M^{(3)} + M^{(4)} + M^{(5)} + M^{(6)} + M^{(7)} + M^{(8)}\right) = \frac{8\rho \, abc}{27}$$

$$\tag{11.37a}$$

$$m_2 = \frac{1}{8}\left(M^{(1)} + M^{(2)} - M^{(3)} - M^{(4)} + M^{(5)} + M^{(6)} - M^{(7)} - M^{(8)}\right) = -\frac{2\rho\ abc}{27}$$

$$(11.37b)$$

$$m_3 = \frac{1}{8}\left(M^{(1)} - M^{(2)} + M^{(3)} - M^{(4)} + M^{(5)} - M^{(6)} + M^{(7)} - M^{(8)}\right) = -\frac{2\rho\ abc}{27}$$

$$(11.37c)$$

$$m_4 = \frac{1}{8}\left(M^{(1)} - M^{(2)} - M^{(3)} + M^{(4)} + M^{(5)} - M^{(6)} - M^{(7)} + M^{(8)}\right) = \frac{2\rho\ abc}{27}$$

$$(11.37d)$$

$$m_5 = \frac{1}{8}\left(M^{(1)} + M^{(2)} + M^{(3)} + M^{(4)} - M^{(5)} - M^{(6)} - M^{(7)} - M^{(8)}\right) = -\frac{\rho\ abc}{27}$$

$$(11.37e)$$

$$m_6 = \frac{1}{8}\left(M^{(1)} + M^{(2)} - M^{(3)} - M^{(4)} - M^{(5)} - M^{(6)} + M^{(7)} + M^{(8)}\right) = \frac{4\rho\ abc}{27}$$

$$(11.37f)$$

$$m_7 = \frac{1}{8}\left(M^{(1)} - M^{(2)} + M^{(3)} - M^{(4)} - M^{(5)} + M^{(6)} - M^{(7)} + M^{(8)}\right) = \frac{4\rho\ abc}{27}$$

$$(11.37g)$$

$$m_8 = \frac{1}{8}\left(M^{(1)} - M^{(2)} - M^{(3)} + M^{(4)} - M^{(5)} + M^{(6)} + M^{(7)} - M^{(8)}\right) = -\frac{4\rho\ abc}{27}$$

$$(11.37h)$$

The matrices of Equations (11.30) to (11.32), with elements as defined in Equations (11.33) to (11.37), are, of course, based on the group-theoretic system of node numbering and positive directions for nodal quantities, as depicted in Figures 11.1 to 11.3, which in general differs from the conventional system. The values are similar, but are not arranged in quite the same way as in conventional matrices.

To render these final results exactly coincident with the element consistent mass matrices based on conventional node numbering and freedom directions, all that requires to be done is to compare the group-theoretic system for the element in question (as depicted in Figures 11.1 to 11.3) with the conventional system and, where necessary, to interchange nodes and reverse the directions of freedoms until the two systems coincide.

Corresponding operations on the rows and columns of the matrices of Equations (11.30) to (11.32) then transform the results into the required conventional form. These last steps, being merely of a renumbering/sign reversal nature, are relatively trivial, so we may regard the results represented by Equations (11.30) to (11.32), in conjunction with Equations (11.33) to (11.37), as the consistent mass matrices in conventional form.

The presented formulation is intended to simplify (and hence accelerate) the numerical computation of element matrices. Once these are obtained, the remainder of the FEM steps are the same as for the conventional method.

11.8 CONCLUDING REMARKS

In this chapter, we have developed a group-theoretic formulation for the efficient computation of matrices for symmetric finite elements. The key feature of the procedure is the decomposition of the displacement field of the element into subfields of the same symmetry types as the subspaces of the symmetry group associated with the element. This leads to the derivation of relatively simple symmetry-adapted shape functions for these subspaces. Symmetry-adapted element matrices (such as stiffness and consistent mass matrices) are obtained separately for each subspace through the integration of much simpler quantities (in comparison with the conventional approach).

Each step of the procedure has been illustrated through derivation of the relevant quantities for the two-node 2 d.o.f. truss element, the two-node 4 d.o.f. beam element, the four-node 8 d.o.f. rectangular plane–stress element, the four-node 12 d.o.f. rectangular plate-bending element and the eight-node 24 d.o.f. rectangular hexahedral element. For these five elements, subspace consistent mass matrices have been derived, and then linearly combined to give element consistent mass matrices, which are the real matrices in the full vector space of the original problem, but based on the group-theoretic pattern of node numbering and positive directions of freedoms. Adapting the results to match conventional numbering of nodes and choice of positive directions of freedoms is readily effected via relatively trivial matrix operations.

Overall, the subspace formulation for symmetric finite elements results in considerable reductions in computational effort in comparison with the conventional approach. The decomposition feature of the method may possibly allow the use of parallel processors, making the computations even faster. However, the formulation only becomes really advantageous in the case of finite elements with high-order symmetry (typically solid hexahedral elements), and a large number of nodes and nodal degrees of freedom. This would then justify the cost of the decomposition inherent in the procedure.

REFERENCES

1. G.M. Zlokovic. *Group Supermatrices in Finite Element Analysis*. Ellis Horwood, Chichester, 1992.
2. A. Zingoni. A Group-Theoretic Formulation for Symmetric Finite Elements. *Finite Elements in Analysis and Design*, 2005, **41**(6), 615–635.
3. M. Hamermesh. *Group Theory and Its Application to Physical Problems*. Pergamon Press, Oxford, 1962.
4. G.M. Zlokovic. *Group Theory and G-Vector Spaces in Structural Analysis*. Ellis Horwood, Chichester, 1989.

Index